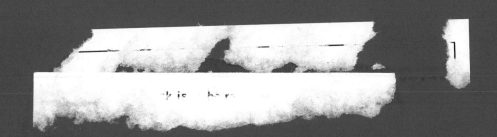

NATURAL LANDSCAPES OF BRITAIN

FROM THE AIR

EDITED BY NICHOLAS STEPHENS

Natural Landscapes of Britain

from the air

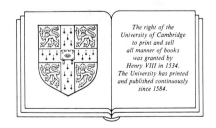

The right of the
University of Cambridge
to print and sell
all manner of books
was granted by
Henry VIII in 1534.
The University has printed
and published continuously
since 1584.

CAMBRIDGE UNIVERSITY PRESS

CAMBRIDGE

NEW YORK   PORT CHESTER

MELBOURNE   SYDNEY

Published by the Press Syndicate of the University of Cambridge
The Pitt Building, Trumpington Street, Cambridge CB2 1RP
40 West 20th Street, New York, NY 10011, USA
10 Stamford Road, Oakleigh, Melbourne 3166, Australia

First published 1990

Filmset and printed in Great Britain by BAS Printers Limited,
Over Wallop, Hampshire

British Library cataloguing in publication data

Natural landscapes of Britain from the air. —
(Cambridge air surveys).
1. Great Britain. Landforms. Evolution
I. Stephens, Nicholas, *1926–*
551.4′0941

Library of Congress cataloging in publication data

Natural landscapes of Britain from the air/
edited by Nicholas Stephens.
     p.   cm.
1. Landforms – Great Britain.
I. Stephens, Nicholas.
CB436.C7N38   1990
914.102–dc20   89-22328 CIP

ISBN 0 521 32390 8 hardback

Frontispiece: Series of meanders of the River Leach,
Gloucestershire, near East Leach (Cambridge University
Committee for Aerial Photography, ACV 15)

Contents

Preface

This book owes its inspiration to a small group of people in the University of Cambridge, where the Air Photograph Library is based, and at Cambridge University Press. The general editors, Robin Glasscock and David Wilson, have provided the initial impetus, while the staff of the Library have supported the search for and selection of photographs in a professional and most helpful way. Almost all the photographs are drawn from the Cambridge collection which owes its origin to the genius of Dr J. K. St Joseph, and currently to David Wilson and his colleagues. Peter Richards and Christine Matthews of the Cambridge University Press and Con Coroneos have provided valuable advice and encouragement at all stages in the preparation of this volume.

The editor has been responsible for the organisation of the book, arrangement of chapter titles and the selection of the contributors. The preparation of each chapter has been designed to examine a wide variety of landscapes and landforms, but the process of selecting photographs from the vast Cambridge collection remained the responsibility of each author. Although an attempt has been made to achieve a reasonable geographical distribution of photographs across Britain for the book as a whole, it will be appreciated that certain landforms are concentrated in restricted areas of the country, as is the case with glacial cirques (chapter 2). There is also the important factor of personal detailed knowledge of some areas at the expense of others, and finally, the availability of photographs of suitable subject material in the Cambridge collection. Inevitably, the process of selection of photographs has meant the exclusion of some topics and the non-representation of large areas of the country, this being but a reflection of the very wide range of landforms with which Britain is so richly endowed.

The companion volumes in this series making up an 'Aerial Survey of Britain' will take up the story of the evolution of British landscapes where the geomorphologist stops. But it is hoped that this series of illustrated essays will provide a useful introduction to the physical framework, and the most important processes which have affected the landscapes and landforms of Britain.

I wish to record my thanks to all those who have contributed to the production of this volume, including especially Sarah J. O'Donnell of the Cambridge Air Photo Library for all her help in the selection of photographs. The patience and skill of Mrs R. M. Morgan (Senior Secretary) is gratefully acknowledged for the production of the bulk of the manuscript; she was assisted at times by Irene Bellinger and Mrs E. Baker; Mr G. Lewis (Cartography) and Mr A. Cutliffe (Photography) provided much valuable assistance; all of the Department of Geography, University College of Swansea. Some figures are based on Ordnance Survey Maps, with the permission of the Controller of HMSO, Crown Copyright Reserved.

Nicholas Stephens

Introduction

There can be few physical geographers studying the evolution of various landforms on the earth's surface who have not consulted and recognised the value of using oblique and vertical aerial photographs to aid their investigations. Although the all-seeing air camera has its own way of recording the landscape, requiring certain skills of interpretation, especially for vertical air photographs, the mobility of a camera carried in an aircraft provides a means of recording the surface features and patterns of landscape below in a way impossible for the earth-bound observer. All the earth-science disciplines have adopted this technique, with geographical studies a principal beneficiary because of the special way in which the ground surface can be viewed. Existing maps can be revised quickly and accurately with the aid of air photographs, while minute detail of the physical, historical and cultural landscape can be recorded at precise times.

However, while air photography alone can provide valuable data about the surface features it cannot replace completely observations on the ground. The correct interpretation of the photographic record depends to a considerable degree upon the experience gained by painstaking field mapping and measurement, the consultation of maps and the examination of documents. Thus, while this volume is devoted essentially to examining a wide range of landforms and landscape-forming processes using air photographs, these are necessarily supplemented by maps, line-drawings and field sketches, as well as information derived from ground-based observations.

Although the total area of Britain is small in comparison with continental Europe the diversity and complexity of its landforms are remarkable, reflecting a highly complex mosaic of rock types and an equally complicated and intriguing geomorphological history. Because of the geological complexity and the variety of landform types present the book has been divided into seven chapters. Each chapter considers a series of examples of landscapes and landforms which have evolved as a result of the interaction between the geological framework and geomorphological processes ranging from ice-sheets to rivers and wave action, and man's activities as a geomorphological agent. However, the book is not designed to be a *systematic textbook* on the physical landscape of Britain; rather it seeks to provide an introduction to the many and varied landform types which can be observed and where at least some informed explanation can be presented to account for the geomorphological features recorded in the air photographs.

The photographs chosen to illustrate chapter 1 demonstrate the major rock formations which make up the geological framework of the country, and illustrate the effects of rock and structure in influencing landform development over many millions of years. The spectacular effects of ice and glacial meltwater erosion are displayed in chapter 2, while the topographic expression of glacial and fluvioglacial deposits is examined in chapter 3. Subtle changes of tone and texture of the ground surface, as recorded by air photography, allows the identification of former periglacial or frozen ground conditions (chapter 4). These air photographs have been deliberately supplemented by a few 'ground' photographs so as to illustrate particular phenomena which are involved in the creation of such surface patterns, or 'patterned ground', to use the technical term. Air photographs can therefore be used to record a variety of relic features of the geological past, formed anything from millions to a few thousand years ago. They can also record the rate of change taking place over much shorter time periods, ranging from a few minutes to a few centuries. Some of the fluvial and coastal features described in chapters 5 and 6 indicate the dynamic and changing nature of certain landforms, for example, where river channels alter position within their floodplains, or where marine sediments accumulate to form beaches and salt marshes. Air photographs have now existed for well over 70 years and they therefore constitute a special kind of historical document. It is clear too that repeated photography can often be used to measure quantitatively the changes of form, and position, of fluvial and marine features, such as river meanders and wave-constructed spits over selected time intervals.

The final chapter seeks to examine some of the changing landscapes of Britain where rapidly operating catastrophic processes have produced major landforms, or have been responsible for significant short-term changes, for example, coastal landslips or river flooding as a result of rapid snow melt. However, the role of man as a geomorphological agent is also considered, where his activities have induced natural processes to operate at increased rates. For example, it has long been recognised that considerable erosion of soils and coastal sand dunes is the direct result of man-made pressure on sensitive environments. Quarrying operations and even severe footpath erosion in some of our National Parks also qualify for consideration as man-made processes responsible for landscape change.

The examination, recording and understanding of the evolution of the natural environment is therefore assisted to a considerable degree by air photography, which in this volume is devoted to the description and analysis of a variety of landscapes and landforms. Some photographs record landscapes that appear to be unchanged and permanent even after aeons of geological time, while elsewhere in a river valley or at the coast there is clearly a highly dynamic and mobile situation. It is hoped that this selection of photographs, together with the accompanying maps and text, will convey to the reader something of the beauty as well as the diversity of the landforms found in Britain, and bring an awareness of the natural forces largely responsible for their evolution.

1 Geology and landforms

Within the relatively small space of Britain there is a wonderful variety of natural scenery: the rugged and often spectacular mountains of Snowdonia, the Lake District, and many parts of the Scottish Highlands; the broad plateau-moorlands of the Pennines, mid-Wales, the south Wales coalfield, Exmoor, Dartmoor and Bodmin Moor; the largely featureless coastal plateaux of Anglesey, Dyfed and Cornwall; the limestone and sandstone cuestas running across England from Devon to Yorkshire; the rounded and distinctive downlands of the chalk of central-south and south-east England; the broad and sometimes ill-drained clay-vales excavated by rivers such as the upper Thames, the lower Severn and the Trent; the monotonous till-plains of Norfolk and Suffolk; and the lowland heaths of the Hampshire and western London Basins.

The form of any physical landscape reflects the interaction, usually over long periods of time, between internal (endogenetic) and external (exogenetic) factors. Internal factors include rock-type and fold- and fault-structures; among external factors, climate is dominant, exerting an obvious control over weathering and erosional processes. Both in broad outline, and on a more local scale, the landforms of Britain inevitably reflect exogenetic processes and events. An example is the protracted erosion cycles of the early Cainozoic, in the course of which extensive surfaces of low relief were planed across pre-existing rocks and structures at an altitude close to the contemporary sea level, only to be upraised to varying degrees by the earth-movements of the Alpine, or mid-Cainozoic, period (as shown in the table on p. 10). Even more obvious erosional events were the Pleistocene glaciations, when much of Britain was greatly modified by both ice-sheets and valley glaciers (see the table below).

Nevertheless, the endogenetic factors have never been completely subdued. Indeed, the fundamental division between Upland Britain (north and west of the Tees-Exe line) and Lowland Britain is clearly based on rock-type and structure.

ERA	PERIOD	EPOCH	AGE IN MILLIONS OF YEARS
C I O Z O N I A C	Quaternary	Holocene-recent	.01
		Pleistocene	2.5
	Tertiary	Pliocene	7
		Miocene	23
		Oligocene	35
		Eocene	53
		Palaeocene	65

Geology and landforms

Major geological features of Britain and the location of photographs 1–17.

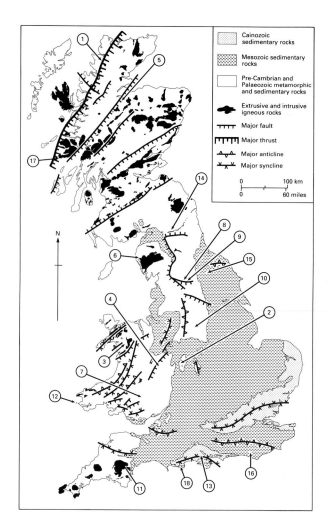

Divisions of geological time.

Age in Millions of Years	Period	Era	Examples of Deposits and areas	Orogenies or Mountain-building processes in Europe	Igneous activity	
					Intrusive (e.g. Granites)	Extrusive (e.g. Volcanic lavas)
	Quaternary	CAINOZOIC	Pleistocene deposits Widespread in Great Britain north of a line London to the Bristol Channel			
50–	Tertiary		Sands and Clays of Hampshire and London Basins	Alpine Orogenesis		
100–	Cretaceous	MESOZOIC	Chalk of the N. and S. Downs			
150–	Jurassic		Cotswold Limestones Lias clays			
200–	Triassic		New Red Sandstone			
250–	Permian	UPPER PALAEOZOIC	Coal Measures Millstone Grit Carboniferous Limestone	Hercynian Orogenesis		
300– 350	Carboniferous					
400	Devonian		Old Red Sandstone	Caledonian Orogenesis		
450–	Silurian	LOWER PALAEOZOIC	Rocks forming much of the Lake District, Southern Uplands and Wales (slates, shales, grits)			
500	Ordovician					
550– 600	Cambrian					
	Torridonian		Sandstones of N.W. Scotland			
	Pre-Cambrian		Schists and Gneisses of Scottish Highlands			

The former is underlain mainly by Palaeozoic rocks which were folded and faulted during the Caledonian and Hercynian mountain-building periods (see table opposite), whilst the latter is underlain largely by Mesozoic and Cainozoic sediments. Moreover, many fine details of the British landscape, from hill-ridges and scarp-edges to individual valley courses, reflect geological control.

In this brief survey of the impact of geology on the British landscape an essentially *chronological* approach will be taken; that is, the rocks and structures will be recounted in order of decreasing age, and their role in landform development illustrated by a series of case-studies. If there is also a broad underlying geographical pattern to the chapter (we shall begin in the north-west of Scotland, and end in south-east England), this will merely reflect the tendency for the rocks and structures to become younger in a south-easterly direction.

THE PRE-CAMBRIAN ROCKS

Underlying Britain is a foundation of ancient pre-Cambrian rocks which is for the most part well concealed. Exposures are usually small-scale and local, occurring where anticlinal uplift and/or faulting have caused them to breach the cover rocks. However, extensive outcrops do exist in the Highlands and Islands of north-west Scotland, where the Lewisian Gneiss, Torridon Sandstone and Moine-Dalradian Schists produce striking landscape elements with no counterparts elsewhere in the country. The Pre-Cambrian rocks of Britain actually vary greatly in age, and also in lithology; in addition to sedimentary and metamorphic types, there are many kinds of extrusive and intrusive igneous rocks (as in western Dyfed and the Malvern area).

Although in detail north-west Scotland has been affected by intense glacial erosion during the Pleistocene, the major landforms are determined by geological structure. The region is underlain mainly by Pre-Cambrian rocks (the oldest in Britain), though to the east of Suilven there is a comparatively narrow zone, of some structural complexity, where Cambrian and early Ordovician sediments (grits, mudstones and limestones) are preserved, as shown in the diagram. These dip to the east, and rest unconformably on the Pre-Cambrian rocks. To the east again lies the great Moine Thrust, marking the termination of the Moine Schists which occupy a large area of the North-West Highlands north of the Great Glen Fault. The Moine Schists (also of Pre-Cambrian age) consist of a great mass of

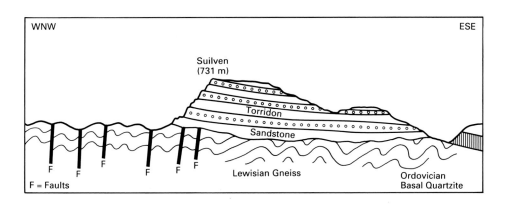

Diagrammatic geological cross-section of Suilven, north-west Scotland.

metamorphosed sediments and igneous intrusions which were thrust far to the north-west, over a low-angled fault-plane, during the Caledonian orogenic phase.

The Pre-Cambrian rocks of the coastal belt are of two types. The Lewisian Gneiss (outcropping over the plateau at the base of Suilven) is a complex of orthogneisses (derived from the metamorphism of plutonic igneous rocks) and paraschists and paragneisses (metamorphosed sediments), together with intrusions of granite and many north-west to south-east trending basic and ultrabasic dykes. The formation is disrupted by numerous faults and shatter-belts, orientated mainly from north-east to south-west. Although the Lewisian Gneiss forms much low lying ground both in Assynt and elsewhere (for example, the Outer Hebrides consist largely of a glacially trimmed lowland of Lewisian Gneiss), it occasionally forms hills and mountains up to 800 m. The irregular terrain and indeterminate drainage pattern of the gneiss – what B. W. Sparks[1] refers to as its 'Laurentian Shield appearance' – owes much in detail to selective erosion by the Pleistocene ice-sheets. Lines of geological weakness have been scoured out, to give innumerable hollows occupied by lochans and marshy ground, and the intervening rises (knocks) show much evidence of plucking and abrasion. However, some elements in the landscape are much older. Prior to the deposition of the second Pre-Cambrian formation, the Torridon Sandstone, the Lewisian Gneiss had been eroded into a landscape of valleys and interfluves which was fossilised by the overlying sediments. More recently erosion has stripped the Torridon Sandstone from wide areas of the gneiss, and has exhumed many ancient valleys. These are in some instances (as in the south of Assynt) being utilised by the modern rivers.

The Torridon Sandstone is a great mass of non-fossiliferous grits and brown feldspathic sandstones exhibiting strong current bedding. At present the formation is near-horizontal, or dipping slightly to the east. However, since the overlying Cambrian strata dip more strongly to the east, at 10–20°, it is evident that the Torridon Sandstone was once tilted to the *west*, but has been modified by post-Cambrian earth-movements. In contrast to the area to the east (and particularly that beyond the Moine Thrust), the Torridon Sandstone shows little evidence of folding or thrust faulting; indeed, together with the underlying basement of Lewisian Gneiss, it appears to have acted as a rigid foreland during the Caledonian orogeny. The occurrence within the sandstone of wind-faceted pebbles, the freshness of the feldspar grains, and the 'torrential' character of the deposit all point to deposition, in areas of shallow water, under desert conditions. At present the Torridon Sandstone gives rise to Suilven and other conspicuous isolated mountains in this part of north-west Scotland. These are characterised by very steep slopes, which in detail often display a stepped appearance owing to selective weathering and erosion of weak strata and bedding planes. In some instances the mountain summits are actually capped by the Basal Grit of the Cambrian formation; this then descends (as at Canisp) the eastern slopes of the mountain as an extended dip-slope.

The Malverns constitute an inlier of Pre-Cambrian rocks, revealed along the 'Malvern Axis' by the removal of younger Palaeozoic and Mesozoic sedimentary rocks. To the east weak Keuper Marls form a gently sloping area (mainly at 20–50 m) descending to the flood-plain of the Severn. To the west predominantly

1 *opposite* The Assynt region of north-west Scotland. The dominant landform is the conical mountain of Suilven (731 m), rising above an undulating plateau surface at 150–300 m. In the left background can be seen the south-western edge of Canisp (846 m). The morphology of this area is typical of much of the coastal belt of north-west Scotland. Suilven is only one of a line of individual mountains which overlook lower ground to the west, and are remnants of a once continuous major escarpment, now broken through by north-west to south-east aligned valleys, many of them containing lochs. NC 160180, looking E.

2 The striking hogback ridge of the Malvern Hills, rising northwards from Chase End Hill (191 m) to Midsummer Hill (286 m) and the Herefordshire Beacon (340 m). Thereafter the 15 km long ridge fluctuates in height before attaining a maximum elevation of 425 m at the Worcestershire Beacon, beyond which it suddenly terminates. SO 762332, looking NW.

Silurian rocks (including the Wenlock Shales, Wenlock Limestone and Lower Ludlow Shales) also give rise to relatively low-lying but morphologically more varied country (including, for example, the narrow Ridgeway, marking the outcrop of the Wenlock Limestone south-westwards from Herefordshire Beacon). Thus although the Malvern Hills are essentially a structural landform, the present-day relief owes much to differential erosion, the igneous and metamorphic rocks of the ridge being far more resistant than the clays, shales and limestones on either side. Finally, it should be noted that the Malverns are one of a series of Pre-Cambrian inliers in the Welsh Border country; others to the north are the prominent hill-masses of The Wrekin, the Long Mynd and Caer Caradoc, in Shropshire, which also owe their prominence to a combination of uplift and differential denudation.

The Pre-Cambrian rocks constituting the Malverns show great complexity though two main categories are recognised. The so-called Malvernian Group – the oldest of all the Border rocks, and probably the equivalent of Pre-Cambrian rocks in Anglesey and northern Scotland – form the central axis of the range. In general terms they comprise schists and gneisses (or, more precisely, diorites with a banded appearance), together with veins of pegmatite (quartz-feldspar rock) and basic intrusions (dolerite). The younger 'Warren House Group', resting unconformably on the Malvernian Group (and, broadly speaking, the equivalent of the Uriconian Group of Shropshire) are mainly of volcanic origin, consisting of lavas poor in silica (spilites), trachytes and tuffs. Their outcrop is limited to a small area at Broad Down, south-east of Herefordshire Beacon.

The present-day form of the Malvern ridge owes much to detailed structural control (see diagram). The axis was initiated by the Hercynian earth-movements of Carboniferous-Permian times, either as an upfold (monocline) whose limbs were

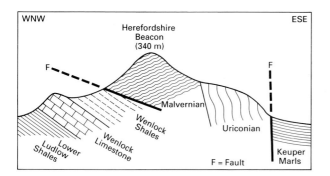

Diagrammatic geological cross-section of the Malvern Hills.

subsequently faulted, or as an upthrust fault-block. The near straight-in-plan eastern margin of the Hills is a fault-line scarp, developed along the major boundary fault (with a downthrow of over 300 m to the east) separating the rocks of the Malvernian Group from the Keuper Marls. It is possible that this scarp (and indeed the Malvern ridge as a whole) is in some measure the product of powerful desert erosion in the post-Hercynian (Permo-Triassic) period. In places the eastern scarp-base is occupied by Haffield Breccia, a deposit laid down by flash floods on an uneven land-surface of Permian date. However, there is no doubt that this ancient scarp (which would also once have been buried beneath other younger rocks, before being exhumed) has been greatly modified, both by more recent faulting and also by erosion, including that associated with the Pleistocene cold periods. The foot-slopes of the Malverns are now effectively masked by superficial deposits, including solifluction debris and gravels transported by meltwater torrents.

The western margins of the Malvern Hills are more complex in structure. At several points there is evidence of powerful thrust faulting, as at Herefordshire Beacon where rocks of the Malvernian Group have overridden to the west underlying Woolhope Limestone and Wenlock Shales. As a result, not only is the western scarp comparatively irregular in outline, but some summits (including Herefordshire Beacon and Chase End Hill) are out of alignment with the remainder of the ridge crest. One interpretation is that the Malvern Hills comprise a series of fault-blocks, differentially thrust to the westwards. The transverse faults between these blocks are lines of weakness which have been etched into well-defined cols (as at Hollybush and Wynds Point) or, in one instance (The Gullet), into a deeply-incised through valley.

THE LOWER PALAEOZOIC SEDIMENTATION

During the lower Palaeozoic (Primary) era, embracing the Cambrian, Ordovician and Silurian periods, vast thicknesses of sediment accumulated in deepening troughs over Scotland, northern England and Wales. This complex basin constituted the lower Palaeozoic geosyncline which, as it developed, was infilled by marine mudstones, shales, sandstones, grits and limestones. Some minor breaks in deposition accompanied temporary local uplifts. In Ordovician times large quantities of volcanic lavas, tuffs and ashes were emitted in an important outburst of extrusive activity, notably in North Wales and the Lake District. These volcanic

3 The northern slopes of Rhinog Fawr (720 m), the highest of the Rhinogs, a small but highly individual group of mountains in north Wales. This is a barren and empty landscape, one of the most striking in the whole of Wales. The exposure of large areas of bare stratified rock, and the etching out by weathering of lines of weakness such as faults and joints, have produced an impression of ruggedness that contrasts with the monotonous plateau-like character of much of the Welsh uplands. SH 650300, looking SE.

rocks, mainly very resistant to denudation, now form some of Britain's most spectacular mountain scenery.

Although the principal characteristic of the North Wales landscape is its highly accidented, even confused relief, there is an underlying pattern which is determined by lithology and structure. Rhinog Fawr and the neighbouring Rhinog Fach (711 m) actually form parts of a discontinuous line of summits (including Craig Wion to the north, and Craif y Ganllwyd to the south-east) which curves round the broad depression, at 200–300 m, drained by the right-bank tributaries of the Afon Eden. These isolated hills are the remaining parts of a cuesta, developed in strata which dip westwards, southwards and eastwards on the flanks of the Dol-wen pericline (whose axis runs north-south from Trawsfynydd towards the Mawddach estuary at Bontddu). The weak rocks in the heart of this fold have been eroded by the Eden tributaries to form the Trawsfynydd lowland, as shown in the diagram opposite.

The Dol-wen pericline itself forms part of a much larger structure, the Harlech Dome. The latter has brought to the surface an extensive outcrop of Cambrian rocks which is bounded to the north, east and south by younger Ordovician formations. These have been strengthened by igneous rocks (for example, the acid rhyolites of Snowdon, and the andesites and rhyolites of the Cader Idris escarpment), and form a half-circle of prominent mountain-masses (at 800–1,085 m) curving around the margins of the Harlech Dome. However, within the latter the 'core' of Cambrian rocks is in general far from unresistant; indeed it has been suggested that the Cambrian sediments, both here and to the north of Snowdon

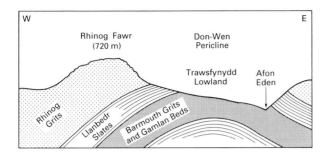

Diagrammatic geological cross-section of Rhinog Fawr and the Trawsfynydd Lowland.

(along the Padarn axis), acted as resistant barriers during the Caledonian earth-movements which produced the main structural elements of this region. As a result the most intense folding occurred along the lines of the Snowdon syncline (between Harlech and Padarn) and the Central Wales syncline (to the south of the Harlech Dome).

The Cambrian strata within the dome consist mainly of flags, shales, slates and – of particular importance – grits. The older sediments are mainly coarse-grained, indicating accumulation on the submerged Pre-Cambrian basement under relatively shallow-water conditions. In detail the Middle and Lower Cambrian (the Harlech Grits) comprise an alternating sequence of grits and shales/slates (in order of decreasing age the Dolwen, Rhinog and Barmouth Grits, separated and capped by the Llanbedr Slates, Manganese Shales and Gamlan Shales). Conditions are favourable for differential erosion of these Cambrian strata, and for the development of a series of inward-facing escarpments (in the hard grits) separated by strike-vales (in the softer slates and shales). However, this simple pattern of landforms has been greatly modified, and in places obscured, by Pleistocene glacial erosion. The most important remaining topographical feature – the disjointed cuesta of the Rhinogs – is formed by the Rhinog Grits. These comprise a maximum thickness of some 750 m of tough and massive greywackes, sandstones and grits with shaly partings. In terms of resistance these contrast with the underlying Llanbedr Slates (200 m, of blue and purple slates), which are exposed at the eastern base of the Rhinog scarp; indeed the relatively rapid erosion of these slates has contributed to the formation of the Trawfynydd lowland, at the heart of the Harlech Dome. The youngest of the Cambrian sediments (finer-grained, deeper water deposits, including the Menevian Shales and Lingula Flags) outcropping on the outer margins of the Dome have also been selectively eroded (for example, along the lower Ffestiniog Vale and the Traeth Bach estuary to the north, and the Mawddach estuary to the south).

The area shown in the photograph comprises a sequence of Silurian sedimentary strata which dip south-eastwards at approximately 10° from the Church Stretton Fault Complex and (immediately to the west again) the upraised pre-Cambrian block of the Long Mynd. The principal individual sedimentary layers are (in order of decreasing age) the Wenlock Shales (mainly grey silty mudstones up to 300 m in thickness); the Wenlock Limestone and associated Limestone Reef Facies (up to 40 m thick); the Lower Ludlow Shales (mudstones and siltstones, up to 250 m in thickness); and the grey argillaceous limestone of the Aymestry Group (up to 60 m thick). The structure, comprising alternating limestones (resistant to erosion owing to mechanical strength and lack of surface drainage) and shales

4 The steep wooded escarpment of Wenlock Edge, its summits rising to 250 m, above a gently sloping scarp-foot zone at 100–150 m. The dipslope of the cuesta declines towards the south-east (left in the photograph) for a distance of only 1–2 km before a second scarp is reached. This is visible in the distance, in the top left-hand corner of the photograph, and in the skyline profile. This 'Aymestry' scarp, unlike Wenlock Edge, tends to increase systematically in height towards the south-west, from 275 m to 339 m. SJ 614010, looking SW.

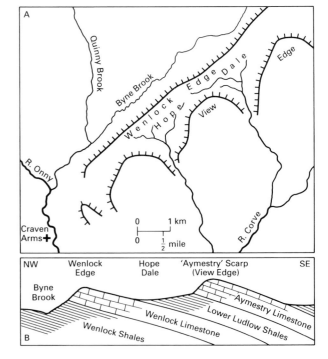

Morphology and drainage of the Wenlock Edge locality (A), and diagrammatic geological cross-section of the Wenlock Edge (B).

18

(unresistant) favours formation of scarp-and-vale scenery (see diagram opposite).

One immediately apparent feature here is the contrast between the continuous nature of the Wenlock scarp (formed by the Wenlock Limestone) and the discontinuous Aymestry scarp (in the Aymestry Group). The latter actually comprises a number of small hill-masses, separated by water gaps marking the passage to the south-east of dip-slope streams which rise in the incipient strike-vale (Hope Dale) etched from the outcrop of the Lower Ludlow Shales. These hills have a distinct prow-like shape, with the apex of each prow pointing up the dip-slope of the Wenlock Limestone. Similar features can be seen elsewhere in Britain, for example the scarp of the Lower Calcareous Grit, overlooking the Oxford Clay vale on the southern slopes of the North York Moors.

One interpretation of discontinuous prow-like scarps is that they are at a relatively early stage of development, involving as yet little integration (by the formation of strike streams along weak outcrops, and resultant stream capture) of the initial dip-slope drainage system. Each individual stream crossing the scarp will lead to the formation of a V-shaped re-entrant (the inverse of a flatiron), owing to the fact that in an up-dip direction the weak stratum (in this instance the Lower Ludlow Shales) lying beneath the cap-rock (the Wenlock Limestone) will rise higher and higher above the stream. The weak outcrop therefore occupies an increasing proportion of the valley-side slope in that direction, promoting more rapid slope recession and the opening-out of the entrance to the through valley. The Wenlock Edge scarp is, however, crossed by only one stream, the Onny at Craven Arms. It is possible that this is the sole representative of a number of streams that once cut through the edge, from north-west to south-east. If such streams did exist, they have been beheaded by the headward extension of the Byne Brook (towards the north-east from the Onny) and the Hughley Brook (south-westwards) parallel to, and at the base of the Wenlock scarp, in the weak Wenlock Shales. This could account for the more advanced development of Wenlock Edge, with its characteristic straight-in-plan scarp. An alternative explanation of the form of the Aymestry cuesta is that a strike stream, occupying the whole of Hope Dale, once flowed at the base of an unbreached scarp-face. Over a period of time this effected lateral down-dip erosion at certain points, resulting initially in the formation of funnel embayments and – at a later stage – the overtopping of the divide at the heads of these embayments and the formation of the present water gaps. This process, identified in the Pennines by A. F. Pitty,[2] has been termed down-dip breaching.

THE CALEDONIAN EARTH-MOVEMENTS

Towards the close of the Silurian period, the formation of red sediments (in the upper Downtonian) denoted a major environmental change from predominantly marine to continental conditions. This resulted from intense folding and uplift associated with the Caledonian orogeny, which fashioned a series of major structures (folds, overthrusts and faults) with a mainly north-east to south-west orientation. This Caledonoid trend is still readily apparent in many major landforms (for example, the Great Glen and Midland Valley of Scotland, and the Lleyn Peninsula of North Wales).

5 The valley of Glen Mor seen from a point 4 km south-west of Inverness. In the middle and far distance lies Loch Ness, occupying what is clearly a glacially modified trough. From the near end of the Loch the river Ness drains north-eastwards through an area of glacial drift which effectively plugs the eastern end of Glen Mor, before entering the sea at the head of the Moray Firth. NH 633415, looking SW.

The main elements in the landscape are the strikingly straight-edged valley of Glen Mor, and the plateau surfaces on either side. Glen Mor in fact extends south-westwards from Inverness, past Lochs Ness, Oich and Lochy to the west coast of Scotland at Fort William and the Firth of Lorn – a total distance of some 100 km. Recent glacial erosion of Glen Mor has been so intense that over much of its length the true valley floor, represented by the deepest parts of the lochs, lies at a depth of more than 200 m below sea-level. The effect is very nearly to make northern Scotland into an island![3] Although also modified to some extent by glacial activity, the plateaus essentially represent much older stages in landscape evolution. That to the right (north) of Glen Mor – truncating siliceous granulite of the Pre-Cambrian Moine Schist Series – lies mainly in the 400–500 m height range, and appears to represent the northerly continuation of the so-called Grampian Lower Surface (at 460–640 m). This upraised peneplain, together with the Grampian Upper Surface (at 730–940 m), was the product of Cainozoic (Tertiary) planation under warmer climatic conditions than exist at present.

The formation of Glen Mor and its lochs has been controlled by the greatest fault-line in Britain, the Great Glen Fault. The numerous important fault-lines of northern Scotland are orientated mainly from west-north-west to east-south-east (for example, that followed by Loch Maree), or from east-north-east to west-south-west (the Caledonoid trend). Many of these faults represent lines of weakness which have existed since pre-Cambrian times (for example, the Loch Maree fault is of Torridonian age), but which have been reactivated several times (but particularly during the main orogenic phases). Within the 'thrust-belt' of the Moine Schists the most important Caledonian fracture is the Strathconan Fault, which has been picked out by a number of small stream valleys. The much more powerful Great Glen Fault is essentially a wrench (transcurrent) fault involving a maximum horizontal displacement of some 105 km (though this is likely to be the net result of several individual movements, some in contrary directions). It is believed that the main dislocations occurred in post-Middle Old Red Sandstone times, though renewed displacement has taken place more recently (for example, that affecting Jurassic strata at the south-western end of the fault-line). Moreover, the occurrence of minor earthquakes shows that adjustments are continuing today.

The formation of the Great Glen Fault has led to the crushing of the rocks on either side of the fracture. Along the south-east margins of Glen Mor a crush-belt affects igneous rocks and crystalline schists, whilst on the northern side (notably between Loch Eil and Lock Arkaig) the country rock is disturbed, though not severely crushed, in a zone 1.5 km in width. The presence of this well-defined line of geological weakness must have influenced denudational processes from an early stage. Thus Glen Mor itself would have been partially excavated by fluvial erosion, following the uplift of the Cainozoic peneplains. During these erosional phases streams became increasingly adjusted to structure, with subsequent streams developing along faults and other lines of geological weakness. It seems likely that, at least in the west, the Great Glen Fault guided one such subsequent stream. It follows that the modification of Glen Mor by Pleistocene glaciation has been merely the final 'event' in a long and complex history of landscape evolution.

In simple terms the Lake District is a large denuded dome, surrounded by a rim of Upper Palaeozoic strata (notably the Carboniferous Limestone) and a central core of older Palaeozoic rocks. The structural evolution of the region has embraced several episodes of orogenic activity, sedimentation and erosion. By far the most important folding occurred in late Silurian–early Devonian times, when the Caledonian earth-movements produced a major anticlinal structure orientated from east-north-east to west-south-west; the axis of this fold is now represented by the outcrop of the Skiddaw Slates (see below). Intrusive igneous activity accompanied the orogeny; in particular a large granite batholith (partially exposed at some points, as at Shap Fell and on the north-eastern flanks of Skiddaw) is now known to underlie the Lake District, and to extend to a depth of 7–10 km. This mass of relatively light rocks, associated with a negative gravity anomaly, has been responsible for continued uplift over a long geological interval.

In the post-Caledonian period erosion on a massive scale, involving the removal of thousands of metres of rock from the Lake District anticline, was followed by a marine episode associated with the formation of a cover of Carboniferous Limestone and – under swamp conditions at a later date – Coal Measures. During the Hercynian orogeny mainly tensile stresses resulted in major faulting in northern England, but further doming of the Lake District seems to have occurred. Subsequently, large-scale erosion again took place during the Post-Hercynian cycle (together with allied deposition of Permo-Triassic strata in peripheral troughs). Any formations of Jurassic or Cretaceous age that were once

developed here have been destroyed, possibly following renewed uplift and warping along an east–west axis in early- and mid-Cainozoic (Alpine) times.

At present the oldest rocks exposed in the Lake District are the Skiddaw Slates, of Lower Ordovician – specifically Arenig and Lower Llanvirn – age. These comprise a complex series of grits, flags, shales and mudstones of marine origin which outcrop extensively in the northern Lakeland Fells. Owing to generally uniform weathering, which gives rise to mantles of shaly debris, the Skiddaw Slates form broadly rounded uplands, as at Saddleback (931 m) and Skiddaw itself (868 m). Following some local foldings, uplift and erosion of the Slates, a great thickness (up to 3,000 m or more) of volcanic rocks accumulated. These comprise the Borrowdale Volcanic Series, of Ordovician – specifically Llanvirn and possibly Llandeilo – date. They now outcrop in a broad belt, to the south of the main fold axis of the Lake District, running from north-east to south-west. Within this belt most of the major summits, such as Scafell Pike (977 m), the Langdale Pikes – visible in the distant left background of the photograph – and Helvellyn (950 m), are developed.

The Borrowdale Volcanics appear to have been extruded onto the floor of the lower Palaeozoic (Caledonian) geosynclinal gulf of north-west England. They consist of a complex assemblage of lavas (mainly andesite, but occasionally rhyolite), fine-grained tuffs, and coarse agglomerates and breccias (which sometimes resemble a crude concrete). In general the Borrowdale Volcanics are highly resistant to weathering and erosion, and have – under the influence of mountain glaciation during the Pleistocene – given rise to a spectacular landscape of crags, cirques, overdeepened valleys and rock basins. However, in detail there is much variation in rock hardness (reflecting contrasts in the degree of induration of the volcanic rocks, and the presence of many faults and joints produced by later earth-movements). The resultant differential denudation has produced – in the words of C. A. M. King[4] – a landscape that is 'knobbly in detail, particularly in areas where the harder members of the series outcrop'. This accounts for the irregular nature of the landscape in general, and – more specifically – for the discontinuous form of Borrowdale itself and the extremely uneven surface of the Thornthwaite–Rosthwaite Fell upland (separating the two upper branches of Borrowdale, in the upper middle part of photograph 6).

There is a further striking contrast in the Lake District between the accidented relief of the Borrowdale Volcanic Series and the low, more subdued Furness Fells region (lying beyond the narrow strip of the Ordovician Coniston Limestone, marking the southern edge of the Volcanics). These Fells are underlain by younger strata (slates, shales and grits of Silurian age).

THE UPPER PALAEOZOIC SEDIMENTATION

Following the Caledonian earth movements, during the Devonian period great thicknesses of Old Red Sandstone (actually red and brown sandstones, pebble-beds and marls laid down in freshwater or terrestrial environments under desert conditions) accumulated over wide areas of Britain to the north of the Bristol Channel. To the south, however, marine conditions persisted, and dull-coloured clays, sandstones and limestones were formed in what is now south-west England.

6 *opposite* The north-trending valley of Borrowdale, in the English Lake District. Although subjected to glacial erosion during the Pleistocene, Borrowdale is far from being a simple glacial trough. Rather it displays both pronounced broadenings (as in the middleground where the Stonethwaite Beck (left) and the river Derwent (right) have filled a former lake with sediment) and constrictions (for example, the narrow wooded valley of the Derwent between Castle Crag and King's How). This irregularity in plan of Borrowdale reflects in large measure the complexity of its geological structure. NY 256186, looking S.

7 The high summits of the Brecon Beacons at Craig Cwm Cynwyn (foreground) and Pen-y-Fan (the table-top hill at 886 m in the left background). The Beacons constitute a magnificent north-facing escarpment of Old Red Sandstone which can be traced westwards to the Carmarthenshire Fans and north-eastwards, beyond the Usk valley, into the Black Mountains. In plan, the scarp face is highly irregular, with a number of obsequent valleys (as at Cwm Cynwyn and Cwm Oergwm) penetrating southwards for 2–3 km. Between these valleys are prominent spurs (such as that visible extending northwards from Pen-y-Fan) which form a dissected scarp bench. The heads of the valleys have been rounded by Pleistocene cirque glaciers, nourished by the shelter afforded by the steep, high northern face of the Beacons. The southern slope is a dip-slope, its bedding-planes truncated and dissected by south-running valleys. In some instances these have been truncated by recession of the Beacons escarpment, producing the pronounced col at 670 m to the east of Pen-y-Fan. SO 030205, looking W.

At present the Old Red Sandstone is best preserved in south and south-east Wales, in central Scotland, and in north-east Scotland and the Orkney Isles, on the sites of the former basins of sedimentation.

In the Carboniferous period (by which time the Caledonian mountains appear to have been effectively planed by erosion), marine conditions were restored over much of the country – though a large land-area continued to extend from central Wales (St George's Land) to East Anglia (the Midland Land Mass). The sediments which formed during the Carboniferous period were, successively, the Carboniferous Limestone (a marine formation), Millstone Grit (a deltaic formation) and the Coal Measures (the product of organic accumulations in muds and sands in a coastal swamp environment). There is, however, local variation in the Carboniferous sequence: for instance, the Midland Valley of Scotland remained a land area during the early part of the period when marine limestones were being deposited to the south.

Structurally, the Old Red Sandstone of the Brecon Beacons forms part of the great east-west synclinorium of the South Wales coalfield: indeed to the south of the coalfield the Old Red Sandstone is again exposed, for example in the Gower Peninsula where it forms hill-masses such as those of Cefn Bryn and Rhossili Down. The structures of the coalfield (which include not only anticlines and synclines, but also numerous faults orientated from east-north-east to west-south-west – such as the great Vale of Neath disturbance, which has been eroded into a deep trench – and from north-north-west to south-south-east) are of Hercynian age. In the vast period of time since the completion of these structures, many thousands of metres of strata have been removed by erosion (some effected in

24

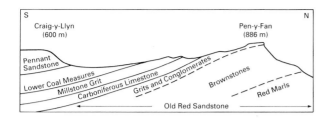

Diagrammatic geological cross-section of the Brecon Beacons escarpment.

the immediate post-Hercynian period, under desert conditions in Permo-Triassic times). Thus the present landscape is, in effect, developed from the roots of the Hercynian fold-mountains.

On the northern margins of the South Wales coalfield, the variety of rock-types, and the prevailing dip (at 5–10°) towards the south, appear to provide ideal conditions for the formation of scarp-and-vale scenery, as the above diagram shows. Other than the Old Red Sandstone, potential scarp-forming rocks are the Carboniferous Limestone, Millstone Grit and Pennant Sandstone (lying above the weak shales and sandstones of the Lower Coal Measures). However, of these only the Pennant Sandstone gives rise to a major scarp comparable in scale with the Beacons. Nor are there major strike vales, followed by well-developed subsequent streams, along the weaker outcrops. Thus the Lower Lime-stone Shales (at the base of the Carboniferous Limestone) form a rather ill-defined depression at the foot of the Beacons dip-slope, and the Lower Coal Measures are marked by an incipient rather than a fully-developed strike vale trending south-westwards from Brynmawr to Merthyr Tydfil and beyond. In fact, the area as a whole remains largely dominated by north–south rivers, which in their course to the south cut discordantly across the fold-axes of the coalfield proper.

The Old Red Sandstone of the Brecon Beacons is not lithologically uniform. The escarpment itself is developed mainly in the Brownstones division, though the actual cap-rock (often forming tabular summits) consists of the tough and resistant topmost grits and conglomerates of the formation. To the north of the scarp, the undulating country (visible in the background of the photograph) extending to Brecon and beyond is formed by the less resistant Red Marls division. However, the morphology of the Beacons reflects erosional history as well as lithology and structure. O. T. Jones[5] has hypothesised that the present scarp is 'inherited' from a Triassic feature produced by desert weathering and erosion. This denudational phase also may have led to the peneplanation of the Palaeozoic rocks over much of mid- and north Wales, and also to the south of the coalfield in Gwent. E. H. Brown,[6] on the other hand, has interpreted the summits of the Beacons as erosion residuals (monadnocks) standing above a mid-Cainozoic peneplain (now represented by plateau-summits at up to 600 m). This surface has been greatly dissected in late-Cainozoic and Pleistocene times, but is still represented by the northward projecting spurs of the Beacons scarp-face.

Giggleswick Scar marks the steep southern margin of the major upland mass of the Askrigg Block. This is mainly a plateau of Carboniferous Limestone (the Great Scar Limestone), which is clearly exposed as a free face (the scar proper) and on the broad summit at the scarp crest (as limestone pavements and other rocky outcrops). Farther to the north the relief of the Askrigg Block is diversified by several high monadnocks, notably Ingleborough, Pen-y-Ghent and Fountains

8a Giggleswick Scar, a 100 m high rocky limestone edge trending south-eastwards towards the valley of the Ribble at Settle; beyond the Ribble rise the southern hills of the Craven District, around Malham. SD 900635, looking SE.

Fell: these are formed in the Yoredale Beds (the uppermost division of the Carboniferous Limestone) and topped by the Millstone Grit. To the south of Giggleswick Scar there is generally low-lying area, at the heads of Ribblesdale and Airedale, eroded from relatively unresistant Carboniferous strata such as the Clitheroe and Pendleside Shales and Limestones and the Bowland Shales.

Giggleswick Scar itself is developed along the western extension of the Middle Craven Fault (one of a series of faults diversifying the structure of the Settle–Malham area), which – with a downthrow of several hundreds of metres to the south – was responsible for juxtaposing the resistant Great Scar Limestone and the weaker shales (see the diagram on p. 27). A long period of erosion during Mesozoic and Cainozoic times removed any younger strata that may have accumu-

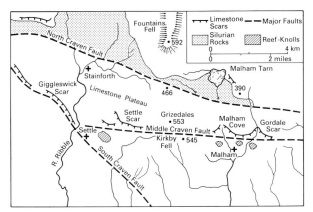

Geological structure and landforms of the Settle–Malham area, Yorkshire.

8b The dry waterfall at Malham Cove. Above the Cove, a dry valley leads north-westwards across the plateau-surface (here at 300–400 m) of Carboniferous or Great Scar Limestone. The former waterfall, which was probably last active during the latter stages of the Pleistocene, was initiated by differential stream erosion along the Middle Craven Fault, which marks the junction of resistant limestone to the north and weaker shales to the south. However, subsequent headward erosion has caused the fall to shift northwards, penetrating the limestone scarp to form the Cove. At its head springs continue to emerge; these are fed in part by an underground stream which enters the limestone by way of a sink-hole a mile or so to the north-west. The stratified nature of the Great Scar Limestone can be clearly seen in the cliffs of the Cove. Above them, differential weathering has resulted in the formation of characteristic 'limestone pavements'. These are near-level areas of rock, coincident with bedding-planes and much pitted by solution along joint lines to give clints (upstanding limestone ribs) and grykes (narrow clefts). SD 900635, looking NW.

lated in the area. The final stages of denudation have involved lowering of the shales, and the formation in the present Scar of an excellent example of a fault-line scarp. It is possible that Giggleswick Scar may also have been emphasised as a relief feature by renewed movement along the ancient fault-line in mid- or late-Cainozoic times, in association with the Alpine movements which so strongly affected south-east England.

Beyond the Ribble valley the fault-line scarp can be traced eastwards, along the Middle Craven Fault, notably at Settle Scar and, in the vicinity of Malham village, at Malham Cove and Goredale Scar. However, in this section it is a more complex morphological feature than at Giggleswick Scar, owing to the occurrence of deeply penetrating valleys (as at Malham Cove), the presence to the south of the fault-line of masses of hard limestone (which have been isolated by erosion of the surrounding shales to form prominent scarp-foot hills), and the preservation at Kirkby Fell of resistant grits which appear as resistant to erosion as the Great Scar Limestone to the north.

The relationships between structure and relief in the Settle–Malham area thus appear to be straightforward. However, for a proper understanding of the present landforms some account must be given of the geological history. During the Carboniferous period the Askrigg Block appears to have been a stable area of limestone deposition, whereas to the south of the Middle Craven Fault – already undergoing development – lay a subsiding trough in which great thicknesses of lithologically distinct strata (including the Bowland Shales) accumulated. However, close to the fault-line irregular masses of limestone (reef knolls) were formed at several points. It was not until Millstone Grit times that similar depositional conditions existed both to north and south of the Middle Craven Fault. This important geological contrast was accentuated by further faulting (in Carboniferous–Permian times) with the creation of the North Craven Fault and the South Craven Fault (branching south-eastwards from the Middle Fault beyond Settle), and further movement along the Middle Craven Fault. The impact of these newer structures remains, however, less marked than that of the older fault-line. Along the North Craven Fault relatively weak Silurian strata (to the north) have been faulted against a downthrown mass of Great Scar Limestone (to the south). Erosion of the former rocks, as yet limited, has initiated what may in time develop into a prominent obsequent fault-line scarp, facing to the north. The South Craven Fault, affecting mainly shales in the Settle area, has little morphological effect.

THE HERCYNIAN EARTH-MOVEMENTS

In late-Carboniferous times the folding, faulting and uplift of the Hercynian (Armorican or Variscan) orogeny began, and continental desert conditions were re-established in the Permian and Triassic periods. The Hercynian structures, whose formation marked the onset of the Mesozoic (Secondary) era, were orientated west–east in southern Britain (for example, in the great synclinorium of the South Wales coalfield, the Mendip anticline, and the tightly-packed folds of the Culm Measures in north Devon), and north–south in northern Britain, as in the Pennine anticline and the major faults marking the western edge of the Alston Block. The Hercynian orogeny also led to renewed earth-movements along

9 The partially wooded eastern slope of the Derwent valley. This rises from approximately 115 m at Curbar village to the plateau of Big Moor (at 300–365 m), itself part of the broad upland known as East Moor. At the summit of the valley slope, and effectively terminating the plateau, is the bold gritstone escarpment of Curbar Edge. SK 245759, looking S.

older Caledonoid trend-lines (such as the boundary faults of the Midland Valley of Scotland). As with the Caledonian folding igneous activity was important. Just as in the former large granite batholiths were emplaced (to form, for example, the Aberdeen–Cairngorms granite), so in the latter the granite bosses of Devon and Cornwall were intruded.

The area shown in photograph 9 lies on the eastern flanks of the great anticlinal structure of the southern Pennines known as the Derbyshire Dome, a product of the Hercynian earth-movements. On either side of the north–south axis of the dome, which brings Carboniferous Limestone to the surface in the Derbyshire Hills, the strata dip both westwards (to the Lancashire coalfield) and eastwards (to the Yorkshire–Derbyshire coalfield). The eastern limb of the fold comprises a sequence of sedimentary strata, including – in order of decreasing age – Millstone Grit, Coal Measures, Magnesian Limestone, Bunter Sandstone and Keuper Marls.

The Derbyshire Dome appears to have been planed across during a major period of erosion (represented by the extensive plateau-like summits of 300–400 m), in the course of which the higher Carboniferous strata were removed from the Pennine axis, and the Carboniferous Limestone exposed and truncated in the Derbyshire Hills. This major planation surface, the so-called Peak District Upland Surface,[7] extends eastwards across the line of the Derwent valley, and is represented by Big Moor (and indeed East Moor as a whole). The age and mode of formation of the surface is much disputed; it has been regarded by some geologists as a sub-Triassic peneplain, formed during the post-Hercynian erosion cycle

10 *opposite* The twin rocky masses of Haytor, near the eastern margins of the Dartmoor plateau. These twin tors reveal a rudely architectural aspect, likened by the geographer D. L. Linton to 'cyclopean masonry'. SX 757770, looking N.

under desert conditions; by others as a Cainozoic (probably post-Miocene) subaerial peneplain; or as the product of marine planation at a level below the Kinder Scout and Bleaklow summits, which rise to over 600 m and preserve remnants of a still higher surface,[8] the Peak District Summit Surface. The latter has been interpreted as part of a pre-Cretaceous erosion surface, subsequently buried beneath the Chalk, but resurrected as a landscape element by Tertiary erosion.

The truncation of the Millstone Grit outcrop by the Upland Surface, both to the west and east of the Derwent valley, provided ideal conditions for differential erosion and the formation of scarpland scenery (see diagram). The Grit (in this locality dipping mainly at a gentle angle towards the east) is lithologically varied, comprising an alternating sequence of relatively unresistant shales and mudstones, and much harder bands of sandstone and coarse gritstones (from which the formation derived its name). Among the latter of particular importance are the Kinderscout Grit (whose outcrop swings round to the north of the Derbyshire Dome, to form a series of steep-edged plateaus and stepped scarps around Kinder Scout and Bleaklow Hill, but which also forms prominent scarps – as at Bamford Edge – to the east of the Derwent) and the Chatsworth or Rivelin Grit (forming a series of north–south edges, as at Stanage Edge, Millstone Edge, Froggatt Edge, Curbar Edge and Baslow Edge). In detail the pattern of the edges is complex. Although most continuous and prominent along the eastern margins of the Derwent valley, gritstone scarps also occur to the west, as at Bradwell Edge and Eyam Edge. The edges form in reality a zigzag pattern, related to minor east–west folding (with the westward projecting salients reflecting small synclines).[9] However, the discordant Derwent has truncated the bases of these salients, isolating the western edges from the main 'line' of scarps to the east of the river. Photograph 9 views Curbar Edge from north to south.

In fine detail, the gritstone edges of this area have been much modified by Pleistocene weathering (in particular frost wedging and associated block disintegration has acted differentially, to produce an irregular scarp-face) and also by quarrying. At the foot of the edges spreads of angular joint-bounded boulders occupy the upper shale slopes. The well-jointed grits are permeable, and below their outcrops spring-lines and seepages occur. These lead to saturation of the underlying shales, rendering them unstable and promoting landslips and (by pressure exerted by the weight of the massive grits) cambering and valley bulging.

Haytor Rocks have been fashioned by denudation from a granite outcrop forming part of the Dartmoor boss (one of five bosses, the others being Bodmin Moor, St Austell, Carn Menellis and Land's End) which trend south of west across the Devon–Cornwall peninsula. Although bearing some imprint of the mid-

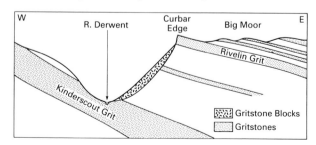

Diagrammatic geological cross-section of Curbar Edge, Derbyshire.

Palaeozoic Caledonian earth-movements, in the form of north-east to south-west structures in pre-Devonian rocks, south-west England was in structural terms essentially the product of Hercynian folding. This produced numerous east–west folds, including a series of domes in the south which, in the later phases of the orogeny, were intruded by large granite cores. In post-Hercynian times massive denudation has stripped away overlying sedimentary rocks to expose the hard granite, which has been left increasingly upstanding as high moorlands, most notably in Dartmoor and Bodmin Moor.

Although the moors, with their generally plateau-like or monotonously undulating character, at first sight appear to display few obvious influences of rock-type and structure, detailed study reveals many subtle but important relationships. For instance, the considerable variation in the form and scale of individual tors owes much to local changes in petrology and the joint pattern of the granite. The latter is, over much of Dartmoor, coarse-grained, with many large porphyritic feldspar crystals. However, there are also younger intrusions of relatively fine-grained, more uniformly crystalline rock (sometimes known as blue granite), together with numerous dykes (for example, the quartz-porphyry known as elvan). In some areas the granite has been profoundly altered by hydrothermal processes (pneumatolysis), resulting in kaolinised or partly kaolinised rock, as at Sheepstor where excavations have revealed 30 m of decomposed *in situ* granite. The jointing of the granite resulted from stresses due to contraction on cooling and from dilatation mechanisms due to denudation and 'unloading'. The granite therefore often displays a cuboidal structure, resulting from the intersection of vertical joints at right-angles and the formation by dilatation of near-horizontal or gently curving sheet-joints. However, there are many variations in joint spacing and type from place to place: for example, the joints of the blue granite are far more numerous than those of the massive porphyritic granite, in which the vertical joints may be 2–4 m apart.

These detailed variations in the Dartmoor granite have largely determined the pattern of denudation. Thus many valleys have been guided either by easily eroded kaolinised granite or zones of closely spaced joints (which have promoted more effective weathering, either by chemical processes such as hydrolysis in a pre-Pleistocene warm and humid climate, or by freeze-thaw action in the Pleistocene). Interfluves and summits are, by contrast, developed in sound or massive granite. However, even on hills and ridges where the most positive relief features (the tors) are formed, differential erosion can be seen. For example, at Haytor itself there are actually two tor masses, each formed by coarse porphyritic granite, as shown in the diagram below. The larger (to the left) is dominated in its upper

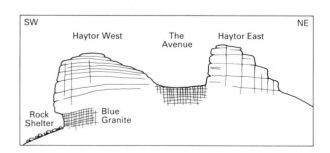

Schematic diagram of the morphological features of Haytor, Dartmoor.

part by sheet jointing, giving it a dome-like summit; it is also strongly asymmetrical, with a steep and high face to the north-west resulting from weathering of a basal outcrop of blue granite. The smaller (to the right) is characterised by massive cuboidal jointing. Between the two is a gap ('the avenue') related to the presence of closely jointed granite. This has been effectively weathered by frost action in the Pleistocene; the resultant debris, comprising numerous small angular blocks, has been spread north-westwards over the slopes bounding Haytor, to give an area of clitter. This surface deposit contrasts with the 2–3 m layer of sandy growan that covers so much of Dartmoor and has resulted from granular disintegration of the crystalline rock.

11 The south-west peninsula and islands of western Dyfed, including Skomer and Midland Island. This area forms part of the county's southern structural region, which is dominated by Hercynian folds. These trend slightly north of west, by contrast with the older Caledonian folds of the northern structural region which trend south of west and influence the general configuration of the north-west peninsula leading to St David's Head. Additionally there are numerous Hercynian faults, including the Ritec Thrust which guided the erosion of the valley system subsequently inundated by the sea to form Milford Haven. SM 710095, looking E.

Much of south-west Dyfed is underlain by Old Red Sandstone, though the outcrop is interrupted by both younger and older formations (for example, at Angle where Carboniferous Limestone is preserved in an east–west syncline and, by influencing differential erosion, has led to the etching out of Angle Bay and West Angle Bay; and at Slatemill Bridge, north of St Ishmaels, where an anticline has led to the exposure of Silurian strata in a narrow east–west outcrop). The most important occurrence of older rocks is in the Marloes peninsula and Skomer. This comprises both Ordovician and Silurian sedimentary rocks, plus the so-called Skomer Volcanic Group (once regarded as of Lower Ordovician age, but now believed by some geologists to be Silurian). The latter consists of an admixture of basalt, rhyolite and dolerite, and in general is highly resistant to denudation.

However, despite the geological complexity of south-west Pembrokeshire, and the presence of Hercynian folds and faults, the landscape inland shows little variety. As photograph 11 illustrates, it is dominantly plateau-like as a consequence of episodes of past planation which have produced in this, and other coastal regions of Wales, a series of coastal platforms, notably at the 200-foot (60 m), 400-foot (120 m) and 600-foot (180 m) levels. Many regard these as upraised wave-cut benches, probably of late Cainozoic (Pliocene) or early-Pleistocene date. In the area shown in the photograph, there is some variability in the height of the plateau. In Skomer and around Marloes it is mainly at 50–60 m, but individual summits rise to 70–80 m; this is indicative either of imperfect planation at the 60 m level, or – perhaps – the presence of a composite surface (it has been suggested that in Dyfed the 60 m platform is a double feature, eroded in relation to former sea-levels at 240-foot (75 m) and 190-foot (57 m) stages). As photograph 11 clearly shows, on Skomer Island the planation has by no means effaced the well-developed east-west structural grain. Several narrow but prominent rocky ridges, marking harder volcanic outcrops, have been developed, possibly as a result of differential wasting of the 60 m platform by subaerial processes since the early-Pleistocene uplift.

Geological structure in this area manifests itself most clearly in its control of marine erosion. The detailed irregularity of the cliff-line reflects differential wave attack on both large and small structures (fault-lines, bedding planes, and soft individual strata). Note the near severance of The Neck at the eastern end of Skomer, where North Haven and South Haven have been eroded along the line of faulted rocks (mainly bedded quartzites). A similar process has formed Little Sound (between The Neck and Midland Island), whilst Martin's Haven and other small bays to the east of Wooltack Point mark the selective erosion of north-west to south-west faults crossing the Marloes peninsula. On Skomer itself, the geo-like inlet of Bull Hole (bottom left of the photograph) is similarly fault-guided.

The Whin Sill is a layer of quartz-dolerite, characterised by a crudely developed polygonal joint pattern, intruded into sediments of the Carboniferous Limestone Series, probably during the early phases of the Hercynian earth-movements. The latter also led to the formation of major north-south fault-lines (the Inner and Middle Pennine Faults), with a downthrow *to the east*, marking the western boundary of the Alston Block. However, in the early post-Hercynian period erosion removed the upraised sediments immediately to the west of the faults, thus forming an obsequent fault-line scarp overlooking the synclinal depression of

12 *opposite* The bold, north-facing escarpment of the Whin Sill between Greenhead and Haltwhistle; in this section the scarp crest, affording a natural defensive line, is followed by Hadrian's Wall. NY 671664, looking E.

The outcrop of the Whin Sill in northern England.

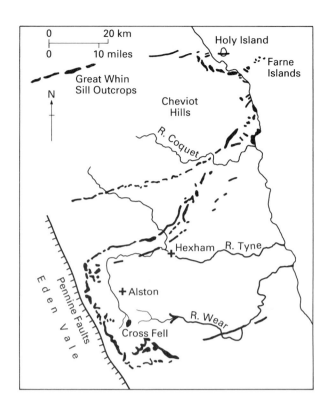

the Eden valley. Within the latter Permo-Triassic sediments (Upper Brockram) contain dolerite pebbles worn from the newly exposed sill (which appears never to have extended to the west of the Inner Pennine Fault).

The Whin Sill is so-called because whin, or whinstone, is a term used by quarry workers to describe dark-coloured igneous rocks, such as dolerite or basalt, which are crushed and used in road construction. The Sill is a major and complex intrusion which can be traced for some 200 km diagonally across northern England from the north Northumberland coast to the western edge of the Pennines at the great Cross Fell escarpment. In general, the dolerite intrusion – which has locally metamorphosed the overlying strata, thus demonstrating that this is not a surface lava flow formed contemporaneously with the sediments – closely follows the bedding of the Carboniferous rocks. Occasionally the Sill (which attains a maximum thickness of 70 m in Weardale, but is on average 25–30 m thick) transgresses from one bedding plane to another, usually by way of a major joint or fault-line. In addition the intrusion sometimes divides into two or more distinct layers; and in some areas it resembles a series of lens-like intrusions rather than a continuous sill of approximately uniform thickness. At places it is also disrupted by small faults of more recent date (as to the east of Greenhead, in the area of photograph 12); these produce slight lateral displacement of the outcrop and the associated scarp.

The Whin Sill does not invariably form a positive relief feature. Whilst in some localities it forms bold crags (as at Hotbank Crags, 8 km north-east of Haltwhistle) resulting from the denudation of adjacent relatively weak strata, in others it has little or no morphological expression. Some of its most spectacular effects are to be seen in Upper Teesdale, where it is widely exposed and gives rise to the High Force and Cauldron Snout waterfalls. It is also much in evidence at the coast between Berwick and Alnmouth. Here it forms crags and headlands (as to the north and south of Embleton Bay, and to the south of Beadnell Bay), and provides sites for famous castles (Lindisfarne on Holy Island, Dunstanburgh, and Bamburgh). It also forms the Farne Islands, which can be viewed from the mainland as 'tilted slabs of Whin Sill dolerite'[10].

THE MESOZOIC AND CAINOZOIC SEDIMENTATIONS, AND THE ALPINE EARTH-MOVEMENTS

During Permo-Triassic times planation of the Hercynian mountains (in the Post-Hercynian erosion cycle) seems to have been completed in many areas. The sediments thus released collected in a series of basins peripheral to the former Hercynian uplands (as in Cheshire, the English Midlands, and a major depression on the present site of the Bristol Channel). It is likely that the Triassic deposits once extended far more widely than at present, to fossilise the Old Triassic erosional land-surface. In some localities the latter has been exhumed to form an element in the present-day landscape (for example, the Vale of Winscombe and the Carboniferous Limestone escarpments of the western Mendips). Some have gone so far as to suggest that the major upland surfaces of areas such as Wales and the southern Pennines also date from this distant period of erosion.

The ensuing Jurassic period was a major episode of sedimentation, mainly under renewed marine conditions, in which great thicknesses of clay, sandstones, and limestone – often in that sequence – were laid down in sedimentary cycles (as in the Kimmeridge Clay, Portland Sand and Portland Stone of the upper Jurassic in southern England). It is not entirely clear how far to the west the Jurassic seas penetrated, though outcrops of Jurassic rocks are found quite extensively in western Scotland. The main area of sedimentation certainly lay in southern, central and north-eastern England; here the varied Jurassic strata form today the basis of a broad tract of scarp-and-vale scenery, with the escarpments being developed in the sandstones and limestones and the vales along the clay outcrops. However, the pattern of landforms is not always simple, for important variations in the thickness and lithology of the Jurassic rocks occur. Some of these are due to contemporaneous uplift along well-defined axial lines (for example, the Mendip and Market Weighton axes). In late-Jurassic times an uplift centred on the oldlands of western Britain caused the sea to regress finally eastwards, so that by lower Cretaceous (Wealden) times only a large freshwater lake remained in southeast England.

By upper Cretaceous times, however, the trend had been reversed, and the sea again spread westwards, across the now eroded surface of the tilted Jurassic and Triassic rocks, in the great Cenomanian transgression (Chalk Sea). In this sea were laid down clays and sands and – far more important – the thick pure white limestone of the Chalk. The great unconformity beneath these upper Cretaceous sediments can be traced in coastal sections in Dorset and Devon.

At the commencement of the Cainozoic (Tertiary) era, uplift of the Chalk (centred on the massifs of the west and north) yet again restricted marine conditions to south-eastern and eastern England. Of particular importance were the

The Cretaceous overstep in
southern England.

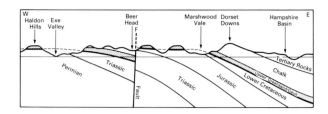

deepening synclines of the London and Hampshire Basins. Whilst the newly
emerged Chalk was being eroded over most of its extent, sands and clays (and
some limestones) accumulated in and around the margins of these basins during
Eocene and Oligocene times. Contemporaneously, a major series of volcanic erup-
tions, associated also with many igneous intrusions in the form of batholiths,
sills, ring-structures and dyke swarms, was occurring in western Scotland. In
south-west England weathered granite and vegetal material from Dartmoor was
being washed eastwards, to collect in the deep trench of the Bovey Basin, south-
west of Exeter. These latter sediments have been correlated with early Cainozoic
(Oligocene) planation on Dartmoor. It is also possible that the upland surfaces
of other parts of western Britain were fashioned by early Cainozoic erosion –
though a strong case has been made for regarding such surfaces in Wales as of
late-Cainozoic age.[6][11] (See the table on p. 9).

The early Cainozoic sedimentation in England was terminated by the so-called
Alpine Storm, which produced many relatively small folds, with some strike faulting, in both Cainozoic and Mesozoic strata. The resultant series of west–east
aligned periclines – many of them breached by subsequent erosion – diversifies
the landscape of the chalk country of Hampshire and the Isle of Wight, Wiltshire
and Dorset. An excellent example of such structural control is provided by the
narrow chalk ridge orientated west-east at Corfe in south Dorset (photograph
13). These fold-structures were evidently completed by a burst of tectonic activity
in early Miocene times, though there is evidence of earlier phases of growth
extending back to pre-Cainozoic times. Elsewhere in Britain the Alpine orogeny
led mainly to gentle flexuring and doming, and a revival of some older faults.
The Miocene and Pliocene periods of the late-Cainozoic are not represented by
significant deposits (other than the shelly sands, or Crags, of coastal East Anglia).

By the close of the Pliocene period the scene had been set for the Quaternary
glaciations, which were to transform in detail the landscape of so much of upland
Britain, and to smother large areas of lowland Britian with glacial tills and glacio-
fluvial sediments (see chapters 2 and 3.)

Morphologically, the Hambleton Hills escarpment – like other Jurassic scarps
in England, such as that of the Cotswold Hills – is quite complex, though at first
sight the photograph reveals most clearly the simple contrast between the 'table-
land' character of the scarp summits and dip-slope and the steep wooded scarp-
edge. In detail, however, the uppermost part of the scarp-face comprises a rocky
free face (at Whitestone Cliff and Roulston Scar), whilst the middle and lower
section is step-like in profile (note, for example, the extensive bench feature at
the scarp base in the left-centre of the photograph). In addition, at some localities
past land-slipping, involving layers of sandstone overlying unstable clay foundations, has modified the scarp form.

13 *opposite* The narrow chalk
hogback, separating the
lowland to the north
(underlain by the weak Eocene
sands and clays of the western
Hampshire Basin) and the vale
to the south eroded from
Wealden sands and clays of
Cretaceous age. The chalk owes
its narrow outcrop, and
relatively restricted elevation
(120–145 m in the foreground)
to its steep angle of dip (up to
80°) to the north, towards the
axis of the major Hampshire
Basin downfold. In effect, the
chalk ridge marks the
boundary between this
syncline and the Purbeck
anticline to the south. The
latter has been extensively
denuded, and the weak
Wealden Beds eroded by the
Corfe river and its tributaries
into a strike-vale. The exit of
the Corfe river from this vale is
marked by the unusual double
water-gap at Corfe, visible in
the middle distance of the
photograph. SY 930820,
looking E.

14 The prominent west-facing escarpment of the Hambleton Hills between Whitestone Cliff (324 m) and Roulston Scar (at approximately 280 m). This scarp marks the boundary between the North York Moors and Cleveland Hills to the east and the lowland of the Vale of Mowbray to the west. The Cleveland Hills constitute a dissected upland block of Jurassic strata, domed along a west-north-west to east-south-east axis by Tertiary earth-movements. The area shown in the photograph lies on the southern flank of the anticline; the strata dip gently (at less than 10°) towards the south and south-east. SE 471849, looking SE.

Geologically, the Hambleton Hills escarpment is developed in Middle and Upper Jurassic strata, with Lower Jurassic (Lias) clays underlying the plain in the foreground of photograph 14. The lower scarp face, and immediate scarp-foot zone, is fashioned from Inferior Oolite (Bajocian) and Great Oolite (Bathonian) sediments, whilst the scarp crest and dip-slope is underlain by Corallian (Oxfordian) Beds; the normally intervening Oxford Clay (Callovian) is greatly attenuated or thins out altogether (as at Roulston Scar). Lithologically the Inferior and Great Oolite of the 'Yorkshire Basin' (separated from the main mass of Jurassic strata by the Market Weighton axis, along which uplift – and hence non-sedimentation – occurred during much of the period) are quite distinct from equivalent deposits in the Midlands and South of England. They comprise mainly deltaic accumulations (rather misleadingly referred to as the Estuarine Series), with marine beds at three distinct levels, attaining an overall thickness of 250 m. The deltaic facies consist mainly of unresistant clays and silts, whilst the intervening marine strata, amounting to some 25 per cent of the total formation, are harder calcareous and ferruginous sandstones, with occasional limestones (for example, the grey argillaceous 'Hydraulic Limestone' which outcrops on the scarp-foot bench beneath Whitestone Cliff). These clay-sandstone alternations are responsible for the terraced nature of the Hambleton Hills scarp.

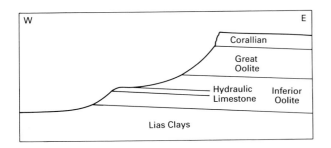

The Corallian Beds, the cap-rock of the escarpment, consist of some 200 m of sandstones (referred to as Calcareous Grit) and oolitic limestones. The so-called Lower Calcareous Grit is of particular geomorphological importance. It not only forms the summital cliffs of the Hambleton Hills, but farther to the east gives rise to a series of prominent flat-iron hills (for example, at Rievaulx Moor, Skiplam Moor and Boonhill Common). This is actually a discontinuous escarpment, similar in some respects to that of the Aymestry Limestone in Shropshire (p. 19), and dissected by streams such as the Rye, Dove and Seven running off the southern flanks of the Cleveland Dome. At the base of this scarp, which constitutes the so-called Tabular Hills, is an incipient strike-vale developed in the weak Oxford Clay (present here). To the north, the moorland summits — underlain mainly by sandstones of the Middle Jurassic — rise to over 400–440 m along the main Cleveland watershed.

The Chalk, of Upper Cretaceous age, is a relatively soft, pure, white limestone with a calcium carbonate content of up to 95 per cent. It is a marine deposit, laid down in a major marine trangression, sometimes termed the Cenomanian transgression, which spread westwards and north-westwards over much of present-day Britain. The initial uplift of the Chalk was associated with some flexuring (the Wealden Dome was initiated by pre-Tertiary folding). However, of greater importance to the structural evolution of south-east England were the so-called Alpine movements, indicated in the diagram on p. 43. These resulted in the formation of a series of anticlines and/or periclines, with a generally east–west orientation and pronounced asymmetry (dips on northern anticlinal limbs often exceed 30°, whilst on the southern they are usually less than 10°). In the South Downs to the north-east of Brighton, Alpine structures include the Falmer Syncline (followed by the valley from Falmer eastwards to Lewes) and the Kingston–Beddingham anticline (breached beyond the Ouse to form the Glynde valley, between the isolated hill-mass of Mount Caburn and the Downs proper). The straight-in-plan form of the escarpment at Ditchling Beacon appears to reflect the powerful influence on erosion exerted by these east–west folds.

In detail the Chalk displays slight but significant lithological variations that have important morphological expression. The division into the Upper (Senonian), Middle (Turonian) and Lower (Cenomanian) Chalks is based on stratigraphy. However, there are contrasts between the relative purity and abundant flint content of the Upper Chalk, and the high marl content and absence of flints in the Lower Chalk (sometimes called the Grey Chalk) which account for differences in resistance to denudation. As a result the chalk scarp is nearly everywhere capped by the lower division of the Upper Chalk (specifically the Echinoid chalk, compris-

ing the *Micraster* fossil-zones), whose resistance is a function of permeability and inability to support surface run-off under present-day conditions. The less permeable Lower Chalk has proved more susceptible to surface erosion, and forms gently concave slopes at the base of the scarp. Within the Upper Chalk there are further lithological variations, sufficient to lead to the formation of a secondary escarpment associated with the pure, flinty chalk of the *Gonioteuthis quadrata* fossil-zone – though this landform is not developed in the area shown by the photograph. However, in this part of the South Downs there is found a prominent bench-feature, marking the outcrop at the base of the main chalk scarp of the hard Upper Greensand.

The steep slope of the South Downs escarpment is indicative of rapid slope recession, promoted by the basal weakness of the Lower Chalk outcrop and by headward erosion by springs. The latter emerge at present at the junction of the Upper Greensand and underlying Gault Clay, but in the past appear to have issued at the outcrop of the Melbourn Rock, separating the Middle and Lower Chalks. It is estimated that scarp retreat during the Pleistocene may have amounted to 2–3 km. Chalk is peculiarly prone to freeze-thaw weathering, and this process, combined with solifluction, has produced the small rounded hollows that diversify the scarp face, perhaps by modifying the heads of spring-sapped valleys. On the South Downs dip-slope the numerous shallow dry valleys, running southwards towards the Falmer valley, also reflect periglacial erosion, by meltwater at a time of permafrost.

Although older rocks occur in Mull (for example, Lower Old Red Sandstone lavas and Mesozoic sediments near Loch Don in the east, and Lewisian Gneiss and Dalradian Schists in the south-west) the geology of the island is overwhelmingly dominated by younger igneous formations. Indeed Mull, together with many of the islands and peninsulas of western Scotland, was in early Cainozoic (Eocene) times the scene of vast outbursts of volcanic activity, embracing the formation of a variety of extrusive and intrusive phenomena. The geological history of this Tertiary Volcanic Region is in detail highly complex, but in simple terms three main phases can be discerned (see tables p. 9 and p. 10).

15 *opposite* A typical section of the chalk escarpment of the South Downs, looking towards Ditchling Beacon. The steep, partially wooded scarp slope (maximum angle 30–35°) to the left contrasts with the much gentler dipslope to the right. Other chalk escarpments in southern England, broadly similar in form and dimensions to the South Downs, include the North Downs (resulting like the South Downs from the erosional breaching of the large anticlinal dome of the Weald), the Chilterns, the Berkshire Downs, and the Dorset Downs. TQ 318138, looking E.

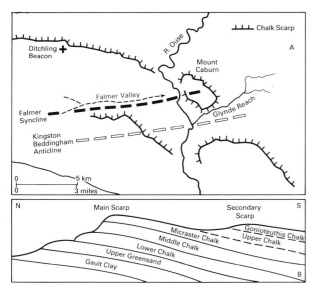

Geological structure, morphology and drainage of the South Downs near Lewes, Sussex.

Diagrammatic geological cross-section of the South Downs.

16 The sea cliff and stepped plateau-edge marking the outcrops of successive basaltic lava-flows at Fionn Aoineach, at the termination of the Ardmeanach peninsula of western Mull. This area is also known as 'The Wilderness'. NM 405273, looking NW.

Initially, large quantities of highly fluid basaltic lava were emitted from an intersecting series of linear vents and volcanoes, to submerge entirely the pre-existing landscape. Over a long period thick accumulations of plateau-lava were built up; in fact, although now reduced in extent and depth by subsequent erosion, some 200 m of lava, covering an area of some 780 km², remain in Mull and account for its predominantly plateau-like character. It is noteworthy that some other Scottish islands, for example Arran, did not experience this initial phase of extrusive activity. To judge from the frequent occurrence of lignites and deep weathered horizons, the volcanic activity was strongly intermittent, with many phases of quiescence or inactivity. The resultant stratified nature of the basalts accounts for the distinctive terraced profiles of many plateau-edges in Mull, including that of The Wilderness.

At a later stage intrusive activity in western Scotland became more active, and led to the development of central intrusions and ring complexes (such as the ring dykes and cone-sheets of the Ben More — Beinn Chaisgildle area of central and eastern Mull, and of Ardnamurchan), many accordant intrusions of the sill type (for example, the quartz-porphyry sill forming Bennan Head and the quartz-dolerite sills giving rise to the plateaus of southern Arran), and much larger scale intrusions (including the gabbro laccolith of the Cuillin Hills of Skye, and the northern granite of Arran).

Finally, large numbers of dykes (mainly of quartz-dolerite or tholeiite) were injected into the existing igneous formations and older sedimentary strata, particularly in Mull, Skye and Arran, resulting in dyke-swarms orientated generally from north-west to south-east. It has been calculated that in south-east Mull 375 individual dykes occur in a belt only 20 km in width, giving an overall thickness of nearly 0.75 km; each dyke averages 1.75 m in breadth. The emplacement of the dykes was associated with a crustal stretching which amounted to 4 per cent in Mull and 6 per cent in Arran (where 525 dykes are developed in a 24 km belt). Differential erosion of these dykes has contributed to the detailed coastal morphology of both islands.

The landscape shown in photograph 17 is typical of much of the Hampshire Basin (particularly in the New Forest and eastern Dorset), and also of the heathy areas of the western London Basin (for example, around Aldershot). The two basins are, in fact, geologically similar, being structural downwarpings initiated by the flexuring of the Chalk in the post-Cretaceous period. This led at first to an important phase of subaerial erosion (the youngest chalk preserved in south-east England is of Senonian age, with no remaining sediments of the Danian and Maestrichtian stages of the Continental Chalk). The resultant erosion surface is preserved beneath the sediments now occupying the basins, and is also represented by the 'sub-Eocene' surface of the lower chalk dip-slopes of the Hampshire Downs, North Downs and Chilterns.

In the Thanetian stage of the early Eocene sedimentation re-commenced in the London Basin (and at a rather later date, the Sparnacian stage — represented by the Reading Beds — in the Hampshire Basin). Throughout the Eocene period a series of marine transgressions and regressions led to the formation of a complex, but generally unresistant, sequence of marine, estuarine and fluviatile sands and clays, reaching a maximum thickness of over 600 m in the northern Isle of Wight. Thus, the London Clay (Ypresian) is of marine origin, whilst the succeeding Bag-

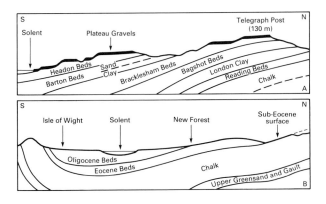

Diagrammatic geological cross-section of the New Forest.

Diagrammatic geological cross-section of the Hampshire Basin.

45

shot Beds (Cuisian) represent the deposits of a river draining from west to east; and the transition from the Middle Barton, through the Upper Barton to the Headon Beds (Bartonian), represents another change from marine to continental conditions. In the London Basin the early Cainozoic sedimentary sequence terminates in the Upper Eocene, whereas in the Hampshire Basin Oligocene sediments (including the Bembridge Limestone) are preserved, particularly in the northern Isle of Wight.

During the Cainozoic sedimentary phase it is likely that earth-movements led to further development and deepening of the London and Hampshire Basins. However, completion of the structures dates from the Miocene period. Subsequently, erosion has been dominant, and it is possible that sediments of late-Oligocene or even early-Miocene age have been removed from the Hampshire Basin. During early- and mid-Pleistocene times, extensive spreads of broken flint gravel (Plateau Gravel) were deposited as a capping on the weak Tertiary rocks in many localities. These appear to have occupied a series of terraces forming a morphological staircase descending from 130 m in the northern New Forest southwards to the Solent shore.[12]

Some terrace levels were evidently of marine origin (formed during periods of high sea level related to episodic withdrawal of the Plio-Pleistocene sea which attained a maximum height of 210 m); others are remnants of river terraces of the Solent River, a former trunk stream (the southern equivalent of the present-day Thames) draining the Hampshire Basin from west to east. In late Pleistocene times the resistant gravel cappings were increasingly dissected by south-draining streams (such as the Beaulieu and Lymington Rivers). In the northern New Forest they have been reduced to narrow interfluve cappings between broad, poorly drained valleys eroded into the weak Eocene sands and clays. In some areas, as at Matley Heath, any previously existing gravels have been entirely removed, and a subdued low-relief landscape formed.

17 *opposite* Matley Heath, on the eastern margins of the New Forest in Hampshire. This is an area of very subdued relief, with broad and gentle interfluves (at up to 30 m) and ill-drained valley bottoms. Running diagonally across the photograph, bottom left to top right, is a well-defined valley bog with a central alder carr. The underlying rock is unresistant, weakly cemented Barton Sand (Upper Barton beds) which characteristically supports acid podsolic soils and – away from valley floors and marshy depressions – dry heathland vegetation dominated by heather and bracken. Beyond the line of the Beaulieu River in the middle distance, the gentle southward dip of the strata results in the exposure of the underlying Barton Clay (Middle and Lower Barton Beds); this supports less acidic brown-earths and woodland becomes more prevalent. SU 330090, looking N.

GUIDE TO FURTHER READING

T. R. Owen, *The Geological Evolution of the British Isles*, Oxford, 1976.

J. G. C. Anderson and T. R. Owen, *The Structure of the British Isles*, Oxford, 1968.

R. J. Small, *The Study of Landforms*, Cambridge, 1972.

B. W. Sparks, *Rocks and Relief*, London, 1971.

2   Landscapes of glacial erosion

Of all the forces that have shaped the landscape of Britain, none has made a more striking contribution than that of moving glacier ice. The role of former glaciers in eroding, transporting and re-depositing rock debris has been debated for more than 150 years. As long ago as 1840, L. Agassiz addressed the British Association for the Advancement of Science meeting in Glasgow on the glacial theory – the idea that there had, in the recent geological past, been an extensive cover of glacial ice over all of northern Europe. In 1860, A. C. Ramsay published a volume entitled *The Old Glaciers of Switzerland and North Wales* in which he correctly inferred the glacial origins of such features as cirques, troughs, erratics and moraines. Three years later, T. F. Jamieson presented a classic paper to the Geological Society of London on the 'parallel roads' of Glen Roy, which were unambiguously shown to be the shorelines of former lakes impounded by glacier ice.[1] On the Pleistocene period in general, James Geikie's *The Great Ice Age* first appeared in 1874.

In some respects, understanding of how glacial landforms were created has not advanced greatly since those years. Although infinitely more is now known about glaciers and the physics of their behaviour, there is still uncertainty about many aspects of the subglacial processes of erosion and deposition: why parts of glacial troughs resemble a parabola in cross-profile, what factors determine the characteristic sequence of basins and rock-bars along a glaciated valley, how many glacial periods there were. These and numerous other important questions remain to be answered.

The selection of photographs depicting the forms of glacial erosion in this chapter is an attempt to convey something of the grandeur of those forms in Britain and their variety, and to show the dramatic imprint of what G. K. Gilbert described as 'the latest and shortest of the geological periods'. In this, the Pleistocene period (the term coined by Charles Lyell in 1839), ice spread repeatedly over the greater part of Britain, extending as far south as northern London in the east and the north coast of Devon and Cornwall in the west. The glaciers were fed primarily by centres of ice accumulation on high ground in the west and north where precipitation of snow was greatest. They radiated out from mountain centres of dispersal in the Cairngorms, Snowdonia, the Lake District and many others, flooding the surrounding country with ice until, during the maxima of the glacial periods, few areas remained uncovered in northern and western Britain. The Irish Sea basin in the west became the collecting ground for a massive body of ice moving south from western Scotland and the Lake District, wrapping around the north Welsh uplands which it was only prevented from invading by the presence of equally powerful Snowdonian ice. On the other side of Britain, Scandinavian ice crossed the North Sea basin to barricade the Scottish ice on the east and divert its flow, while farther south in eastern England, it succeeded in penetrating far inland. The flow of the ice was partly controlled in each glacial period by the relief forms already existing, so that certain preferred routeways were repeatedly emphasised and deepened. In some other areas, the cover of ice became so massive that the existing topography was completely over-whelmed and the ice became able to move, in part, independently of it. D. L. Linton in 1949 was the first to examine in detail such a situation in the Scottish Highlands, where the main ice dome lay east of the mountain backbone and where ice-streams crossed the latter in seeking an Atlantic outlet.[2] In doing so, the ice-streams excavated new valleys as well as deepening older ones.

The maximum extent of the ice in each glaciation differed. The southernmost limit already referred to was not attained in the last main glacial period. The last main glaciation, termed the Devensian (the equivalent of the Weichselian in northern Europe and the Würm in the Alps), culminated about 18,000 years ago, when it reached a southernmost limit along a line from the Vale of York, past Wolverhampton towards south Wales. In Scotland, estimates place the verti-cal extent of this ice at more than 1,800 m over the Scottish Highlands (cf. Ben Nevis 1,343 m) and, in Wales, at up to 1,500 m (cf. Snowdon, 1,085 m).[3] Since an earlier glaciation, the Anglian, had a greater horizontal extent, reaching the south Midlands and East Anglia, we can only conclude that its ice attained still greater thicknesses.

The total number of glaciations to have affected Britain is not known. Certainly there were at least four, including the minor re-advance of the Loch Lomond Stadial, lasting from about 11,300 BP to 10,100 BP, which was separated from the preceding Devensian by a short interval, the Allerød interstadial, when ice probably vanished altogether from Britain. No glaciers have existed in Britain since the Loch Lomond Stadial, though the theoretical snowline does not lie far above the tops of the highest mountains in Scotland today. The features depicted on the photographs have all passed through a long history of Pleistocene evolu-tion, with its glacial and interglacial oscillations. Although the imprint of the Devensian glaciation is the clearest (and in Scotland there is hardly any trace surviving of older glacial or interglacial events), it must not be thought that the present landforms of glacial erosion were created in their entirety at this time. We have no means of determining, however, how the individual valleys or cirques evolved during the older stages, nor what the exact form of the preglacial land-scape was like, but constantly the preglacial, older glacial and interglacial inheritance must be borne in mind.

It is very clear, when studying the spatial distribution of glacial erosion forms, that certain areas were more intensively glaciated than others. An attempt is made in the map (on p. 50) to delineate zones of differing intensity of glacial erosion, based on the presence or absence of certain diagnostic features, or the degree of their development.[4] The explanation of the patterns of zonation is complex, linked to geological, topographical and climatic factors: for instance, the strength of the rock and its competence to stand in steep faces, the amplitude of the relief, and the spatial distribution of precipitation in the glacial periods. Many of the selected photographs are located in zones III and IV, for here the characteristic glacial features are best developed.

The spatial scales of glacial erosion forms vary from the microscale to

Zones of glacial erosion in Britain.

Zone 0   No glacial erosion.
- I   Glacial erosion limited to slight modification.
- II   Extensive glacial excavation along main flow lines; ice-scoured bluffs, crag-and-tail features; glacial troughs.
- III   Preglacial forms considerably modified or even replaced. Comprehensive glacial trough systems.
- IV   Almost complete transformation of the landscape by glacial erosion.

The location of aerial photographs is also shown.

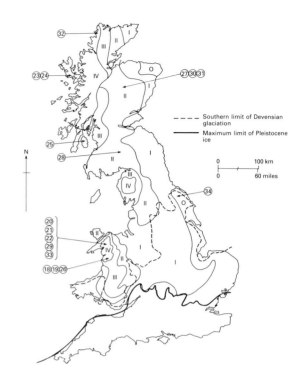

macroscale. Microscale features such as striations and friction cracks cannot be identified on photographs of the scales included here, but have contributed importantly to understanding of glacial erosion processes. Mesoscale features can sometimes be located individually on the photographs but the scale of the latter is more suited to the study of their patterns in groups. The complexity of mesoscale glacial relief is often so great that, for its analysis, air photographs are often superior to the most detailed topographic maps – it is instructive, for example, to compare photograph 32 with the corresponding Ordnance Survey map. The outstanding forms seen on the photographs are of macroscale though, where dimensions are measurable in hundreds or thousands of metres.

It has long been known that some features of glacial erosion show a resemblance to relatively simple geometrical forms – the parabolic cross-profiles of glacial troughs, the arcuate outlines of cirques and so on. The study of the morphometry of glacial features has not been so highly developed as the corresponding study of fluvial forms. Nevertheless, it can yield interesting comparative data, such as trends in spatial distributions, and can offer further questions for investigation in relation to the operation of the formative processes – for instance, *why* do glacial troughs sometimes approximate to a parabola in cross-profile? The following table gives data for four simple cirque parameters to illustrate the value of morphometric studies. The data from which the values have been calculated were obtained from Ordnance Survey maps. Height data on such maps, especially at the higher elevations, are of varying reliability and this must be borne in mind, together with the possibility of subjectivity in measurement; field measurements would, of course, be preferable.[5] (See the table on p. 55.)

The cirques chosen are all shown on the photographs that follow. Length–height ratios calculated according to G. Manley's suggestion[6] yield remarkably similar values, all except one falling within Manley's claimed range of between 2.8:1 and 3.2:1 for well-developed cirques. Two measures of the degree of closure of a cirque, in plan and in profile, are adopted from I. S. Evans.[7] The first is the range of azimuth possessed by the longest contour, the second is the difference between the steepest backwall slope and the minimum outward floor slope (if the cirque floor has a reversed slope, the two angles are added). Over much larger populations, trends in such variables, if the values are not random, demand explanation. The fourth variable, the maximum possible slope of a glacier occupying the cirque, should be considered in the context of W. V. Lewis's claim[8] that, where such angles exceed 7° or 8°, rotational sliding of the glacier and hence scouring of the cirque floor to a basin-like form will begin to occur.

To an extent, then, inferences can be drawn about form-process relationships from photographic and cartographic images, but only of a rather general nature. The mechanics of the processes involved in the sculpturing of glacial erosion forms continue to present many problems, principally because of the difficulty of investigating such processes beneath present-day glaciers. One point that is becoming very clear is that the separation of glacial from glacifluvial processes is often extremely arbitrary, both sets of processes working together, at least under warm-based glaciers. In any given cross-section of a glacier, far more material may be in transport by subglacial streams than by the ice itself, and the erosional capabilities of such streams can also be immense. Two of the photographs have therefore been selected with this theme in mind, one (34) showing a classical meltwater channel in north Yorkshire, the interpretation of which has changed radically since the turn of the century, and the other (33) of a familiar assemblage of features in a glacial trough – a rock-bar or valley-step cut through by a deep V-shaped gorge. In the second case, the cutting of a gorge through a rock-bar has often been attributed to subglacial meltwater erosion. Since it is impossible that such gorges were formed wholly or even partly in the post-glacial, the gorge must have existed under (and been partly filled by) glacial ice, leading one to suspect that its cutting was due partly at least to the concentrated flow of subglacial meltwater in this position.

The sequence of photographs begins with cirques, the most distinctive of all glacial erosion forms at the macroscale. Cirques are illustrated both as individual features and as members of a series in a 'cirque staircase'. Enlargement of adjacent cirques produces an arête (German: Grat) between them, and several photographs show the typical pyramidal peaks (or horns) standing at the junction of three or more arêtes. Glacial troughs are the subject of the next group of photographs, exemplifying the characteristic cross-profiles, rock-basins, valley-steps and trough-ends. The theme of the third group is the phenomenon of glacial diffluence and transfluence, where ice seeking new outlets at the maximum stages of glaciation sometimes transgresses former interfluves, in some circumstances completely breaching them. One photograph (32) is included to demonstrate ice erosion in an area of low relief. The final pair of photographs, as already noted, illustrates the theme of meltwater erosion.

Cwm Cau, containing the lake Llyn Cau in the centre of photograph 18, is the most southerly well-developed cirque in Britain. It has been described as one of the finest examples of the cirque form, depicting a complete range of fea-

18 Cwm Cau and Cader Idris. The summit of Cader Idris near the top is pinpointed by the convergence of footpaths. In the centre of the photograph is the cirque of Cwm Cau, occupied by the rock-basin lake of Llyn Cau. The stream from this lake drains east before turning to fall steeply towards the Tal-y-llyn trough, part of whose floor is seen at the lower edge of the photograph. Part of the cirque containing Llyn-y-Gadair is visible at the top left. SH 71514.

tures associated with this type of glacial sculpture. To obtain the best appreciation of its form and its relationships with the surrounding landscape, photographs 18 (vertical) and 19 (oblique) should be studied together.

The surface of Llyn Cau stands at 467 m, 425 m below the summit of Cader Idris (identifiable most easily on photograph 19) north of the cirque. The main axis of the cirque runs east-north-east, but beyond the cirque threshold about 100 m east of the point where the stream leaves the lake (scale may be assessed from the length of the lake, about 600 m), the valley gradually swings round to a southerly orientation, before falling steeply (a typical hanging valley) to the floor of the Tal-y-llyn valley at the lower right corner of photograph 18. The lake occupies a true rock basin, retained by the threshold in solid rock, though it is possible to see on the photographs some of the small moraine ridges dating from the Loch Lomond Stadial that cap the threshold and occupy a small area between the threshold and the lake.

The walls of the cirque are impressively steep, the overall angle of slope reaching 54° and even vertical for short pitches. The steepest part, picked out by the shadow on both photographs, is the headwall south-west of the lake. This corresponds to the outcrop of the Upper Acid Group of the Ordovician Llandeilo Series (see accompanying map, p. 54), these tough igneous rocks being competent

to stand in such a high face. Farther north, the angle of slope of the cirque wall becomes noticeably less, corresponding to the outcrop of Llyn Cau mudstones. The Ordovician rocks dip steeply (about 45°) to the south, the strike following the axis of the cirque and presumably controlling its alignment. It is suggested that, preglacially, an eastward-trending valley was eroded by a river picking out the softer Llyn Cau mudstones before glacial modification began. Beneath the shales, resistant sills of granophyre and dolerite rise to form the summit ridge of Cader Idris with its north-facing escarpment. Into this escarpment face, other cirques have been carved, notably that containing Llyn-y-Gadair visible at the top of photograph 18.

The cirques of the Cader Idris massif face directions between north and east, reflecting climatic control of cirque development. During glaciation, north- and east-facing slopes in the lee of the prevailing snow-bearing winds and in shade for the longest time provided the most favourable locations for snow accumulation

19 Two overlapping views of Cwm Cau and the summit of Cader Idris, looking generally northwards. SH 711131.

The relief and geology of the Cader Idris area.
Geology: 1. Tal-y-llyn mudstones
          2. Upper Acid group
          3. Llyn Cau mudstones
          4. Upper Basic group
          5. Llyn-y-Gadair mudstones
          6. Lower Basic group
          D Dolerite
          G Granophyre

and cirque glacier nourishment. The powerful control exerted on cirque aspect and development can be vividly seen by comparing (photograph 18) Cwm Cau with the head of the valley draining in the reverse direction towards the left-hand edge of the photograph. The two valley-heads stand back-to-back, developed in the same geological formations, but the head of the east-facing one has been sculptured into the massive cirque of Cwm Cau while the west-facing head is that of a normal river valley, with comparatively gentle slopes and showing few signs of glacial modification.

The whole area, with the possible exception of the summit ridge, was submerged by ice during the Pleistocene glacial maxima. At other times of less complete ice cover, however, the glaciers in Cwm Cau, the Llyn-y-Gadair cirque and so on, were able to continue to erode their basins strongly. At times when Cwm Cau supported its own separate cirque glacier, the maximum angle attainable by the ice surface would have been 21°, measured from the highest point of the backwall to the lip of the cirque threshold. This comfortably exceeds the minimum angle necessary for rotational sliding of the ice in a cirque basin to occur. This suggests that the cirque glacier in Cwm Cau was highly active, eroding the rock basin in its floor to give its present well-developed cirque form. The floor of the cirque and of the valley continuing to the east shows numerous roches moutonnées and glacially abraded rock surfaces, contrasting with the shattered appearance of the headwalls often ascribed to freeze–thaw processes.

Photograph 19 shows how the cirque is abruptly cut into a high-level surface, declining gently to the left and to the right of Cader Idris summit. Over the area of the photograph, this surface ranges in height between about 700 m and 893 m, and probably represents the remnants of the preglacial (? pre-Pleistocene) topography. Into this surface, river valleys were first incised, and later modified and deepened by ice during successive glacial epochs. It is not possible to determine how much of the excavation of Cwm Cau is attributable to any individual glaciation, nor how much to other processes of erosion in interglacial periods. The last ice in Cwm Cau existed in the Loch Lomond phase, a short-lived glacial episode spanning a few centuries around 10–11,000 BP, when the snowline must have descended to about 650 m.

Some variables in cirque morphometry

CIRQUE	PHOTO NUMBER	LENGTH— HEIGHT RATIO	CLOSURE IN PLAN (DEGREES)	CLOSURE IN PROFILE (DEGREES)	MAXIMUM GLACIER SLOPE(S) (DEGREES)
Cwm Cau	18	2.8	195	73	21
Cwm Glaslyn	21	3.0	210	54	17
Cwm Du	20	2.0	163	54	19
Coir-a-Ghrunnda	23	2.9	202	55	19

Photograph 20 shows a cirque in north-west Snowdonia made famous in the geomorphological literature by W. M. Davis's description and sketch of it in his paper of 1909 on 'Glacial erosion in North Wales'.[9] The name of the cirque is Cwm Du, the 'black hollow', and it is cut into the mountain Mynydd Mawr

20 The cirque cut into the northern face of the round-topped hill of Mynydd Mawr, Snowdonia. In the far left distance may be seen part of the upper valley of the Colwyn. Also in the distance is the line of hills from Y Garn to Mynydd Tal-y-Mignedd, scalloped by north-facing cirques. SH 540547.

(698 m): Davis wrote of how 'Cwm Du breaks the arch of Mynydd Mawr as abruptly as if it were a huge quarry'. The cirque is beautifully proportioned in plan, completely symmetrical about an axis trending slightly west of north, but it lacks the degree of closure (see the table above) characteristic of the cirques shown on photographs 18, 19 and 21 – that is, its sidewalls actually diverge (by about 17°) rather than enclose the hollow. Furthermore, it lacks the basin-shaped floor of a fully-developed cirque, the floor here sloping outwards about 5° though there is beneath it a small but unknown thickness of drift and colluvium. The symmetry of the cirque is partly a reflection of a relatively uniform lithology, the Cambrian slate outcrop dipping away from the observer towards the south-east at about 30°.

The strikingly smooth and rounded summit of Mynydd Mawr has been claimed as a remnant of the preglacial land surface, and there is no reason to dispute this view, though a two-stage model of development such as Davis proposed (preglacial and glacial) is much too simple. The cirque must have evolved through a sequence of glacial erosion phases separated by interglacial epochs of weathering, mass movement and water erosion. The latest phase in its evolution was that of the Loch Lomond stage when a small glacier re-occupied the cirque, though it failed to develop any significant moraines. The distant view includes a row of peaks from Y Garn to Mynydd Tal-y-mignedd on the right, rising to between 630 and 710 m, possibly belonging also to Davis's summit surface. Their northern faces also bear cirques with orientations similar to that of Cwm Du.

The east face of Snowdon, whose summit at 1,085 m appears on the skyline near to the left-hand margin of photograph 21 is characterised by a series of cirque basins rising one above the other in a typical cirque staircase. Three such cirques can be distinguished, the lowest containing Llyn Llydaw in the foreground (average water level 430 m), the second containing Llyn Glaslyn (water level 601 m), and the third, much smaller and less distinct, standing about half-way up the steep face rising from Llyn Glaslyn to the summit ridge. Its floor level is at about 750 m, and it lacks the well-developed forms of the lower cirques, especially their rock basins. The rock basin containing Llyn Glaslyn is 39 m deep (measured below the water level); that of Llyn Llydaw is 58 m deep. Neither lake is moraine-dammed, rock appearing in the beds of the streams draining over the cirque thresholds, though late-glacial, Loch Lomond stage moraines are present around the lakes, especially Llyn Llydaw.

The symmetrical morphology of a typical cirque is beautifully displayed by the Glaslyn basin, the enclosing arêtes descending and approaching the lake outlet. It has a length–height ratio of 3:1, ignoring the third minor cirque half-way up the backwall. Its aspect is open to the east, an orientation already encountered for Cwm Cau (photograph 18) and common to many British cirques.

At the maximum of the last (Devensian) main glaciation, the whole series of cirques was occupied by ice pouring out of one basin into another, with only the summit ridge and upper parts of the arêtes visible above the ice surface. During deglaciation, about 15,000 years ago, the ice thinned until the rock bar enclosing Llyn Glaslyn began to be more and more exposed. Thus the ice supply to the lower basin of Llyn Llydaw was gradually cut off. For some time, the separated Llydaw glacier may have been able to survive, and even continue to move, but as the snowline steadily rose, it eventually became climatically dead. The higher Glaslyn glacier was able to survive longer. After complete deglaciation by the Allerød phase, glaciers re-occupied both cirques during the Loch Lomond stage, 10,000–11,000 years ago. The patterns of moraines (only small examples can be picked out on this photograph) suggest that ice for a time again flowed from the Glaslyn to the Llydaw cirque.

Other features discernible on the photograph include the heavily-used hiking paths – the Miners' Track ascending from Llydaw to Glaslyn, and the higher Pyg Track crossing from the middle right to the Snowdon summit ridge. The level of Llyn Llydaw fluctuates, as its shoreline shows, because of its use as a reservoir to supply a small hydro-electric plant.

The imposing ridge of Y Lliwedd (photograph 22) is one of a series radiating from the central peak of Snowdon which lies behind the observer. The view is to the east along this typical example of an arête formed by the erosion of the

21 The series of cirques on the east face of Snowdon. Snowdon summit is on the skyline near the left edge of the photograph, with a prominent ridge descending from the summit towards the observer. The lowest cirque (foreground) contains Llyn Llydaw, the next Llyn Glaslyn, while faintly visible on the slope between Llyn Glaslyn and the summit ridge is the trace of a third incompletely developed cirque. SH 609543.

22 The asymmetrical arête of Lliwedd, seen from a point above Snowdon. On the left is part of the cirque containing Llyn Llydaw (the lake is not visible). SH 625532.

cirques on either side – on the left, the cirque containing Llyn Llydaw (the lake is not visible on this photograph), and on the right, Cwm Llan, from which the Watkin Path can be seen ascending to the lowest point on the Lliwedd ridge, Bwlch Ciliau at 744 m. From Bwlch Ciliau (Bwlch = saddle), the arête climbs steadily to a maximum of 898 m. It is built of Ordovican rhyolitic tuffs, dipping steeply towards the observer.

The contrast in overall slope angle between the two sides of the arête is striking. On the left, the north-facing slope drops sharply at angles between 60° and, locally, vertical, whereas on the right the descent into Cwm Llan is, on average, 30°. This difference is mostly a result of climatic control, the north face providing the most favourable conditions for snow accumulation (as described above,

p. 53). The cirque glacier banked against this wall would therefore survive long after the ice on the south-facing side had melted; in addition, freezing temperatures would be more frequent and prolonged on the north face, as indeed they are at the present day in winter. The col of Bwlch Ciliau marks the point where the two cirques approach one another most closely, this relationship probably reflecting the tendency for dilatation jointing to develop in the bedrock during cirque erosion, the joints forming parallel to the cirque outlines.[10] In arêtes, two cirques would intersect at shallow angles and weaken the ridge, hastening its destruction by glacial erosion and frost weathering.

The Cuillin Hills in the Isle of Skye represent the most concentrated assemblage of spectacular glacial erosion forms in Britain. In an area of about 40 km^2, no less than 24 major cirques can be identified, arêtes totalling 38 km in length, rock basins, glacial troughs, hanging valleys, rock bars and valley steps. In addition, there is no other area in Britain with so great a proportion of bare rock and such steep slopes. The two photographs portray a representative selection of these features, the first (23) of the cirque Coir-a-Ghrunnda and the second (24) of the Coir-uisg, the trough containing Loch Coruisk.

The Cuillins consist geologically of a laccolith of gabbro intruded at depth during the early Tertiary period, some 60 million years ago. Little evidence is available to reconstruct any elements of the Tertiary landscape, save that the main summits of the present-day mountains fall within a narrow range of 900–1,000 m. Many see in this a Gipfelflur, the last vestige of a high plateau that was once bevelled across the exposed gabbro, subsequently uplifted and then deeply dissected before the onset of glaciation. The accordance of summit level is well brought out on photograph 23 and gives a measure of the intensity of late Tertiary and Pleistocene dissection. The sharp peak above the lake and slightly to its left is Sgurr Alasdair, at 993 m the highest in the Cuillins, while photograph 24 shows how, on the other hand, the mountains descend to sea-level. Relief amplitude can be as much as 700 m km^{-2} and slope angles approach the vertical in many localities. On photograph 24, the right-hand (north-east) wall of the Coruisk trough has an average slope of 48° above the far end of the lake, and on photograph 23, the facing wall of Coir-a-Ghrunnda stands at no less than 51°.

In the Pleistocene, the Cuillins supported an independent system of glaciers, massive enough to prevent invasion by the mainland ice of Scotland. The trough of Coruisk was fed by eight or more cirques, together with a glacier from the hanging valley of Coire Riabhach on the right of photograph 24. The main glacier eroded the rock basin now holding Loch Coruisk, whose water level stands at 8 m above sea level and whose floor falls far below it. Separating the lake from the sea is a prominent rock-bar or Riegel cut across by the channel of the Scavaig River. Coir-a-Ghrunnda is a finely proportioned cirque containing a small rock-basin lake at 695 m. Anomalously for most British cirques it faces south-west, but the Cuillin cirques were fed by so much precipitation that they can be found facing in all directions, the only evident climatic control being that the floor levels of those facing south and west are higher than those facing north and east. Coir-a-Ghrunnda stands back-to-back with Coir-an-Lochain, the sharp arête between them dropping to a notch near the centre of photograph 23 at 860 m. The cirque

23 *overleaf* Coir-a-Ghrunnda in the Cuillin Hills, Skye and its lake. Several of the highest peaks of the Cuillins can be seen in this photograph to rise to a common summit level of 900–1,000 m. Also evident is the extent and steepness of many bare rock slopes, and the sharpness of peaks and arêtes. NG 452202.

backwall shows some steep debris chutes and limited accumulation of talus near its foot, but overall the lack of loose rock or moraine is remarkable, the glacial forms showing hardly any sign of modification by weathering or erosion since the ice vacated the region.

The island of Arran in western Scotland shows many fine examples of glacial erosion forms, located in the northern half of the island which consists of a Tertiary granitic intrusion, fine-grained granite in the centre being surrounded by coarser-grained granite, the latter occupying most of the area of the photograph. Both types of granite are competent to maintain steep faces in cirque headwalls, arêtes and peaks, and display frequent close relationships between its joint patterns and land form. The highest point is Goatfell (874 m) to the west of which

24 The glacial trough containing Loch Coruisk, Skye. The lake surface stands 8 m above sea level, connected to the sea by the Scavaig river. The rock-basin floor of the loch descends far below sea level. NG 489197.

25 The glacial U-shaped valley of Glen Rosa, Arran. In the left distance, by the head of the valley, the pyramidal peak of Cir Mhor may be picked out. NR 983387.

lies the glacial trough shown in photograph 25, a classic example of a so-called U-shaped valley. The view is towards the north, up Glen Rosa towards the pyramidal peak (left distance) of Cir Mhor (798 m) which stands at the intersection of three radiating troughs and arêtes. Glen Rosa was fed by ice from cirques on the left (one seen in deep shadow), and ice also appears to have spilled over the divide to the right of Cir Mhor where a broad saddle leads into Glen Sannox. Arêtes flank the valley of Glen Rosa, the right-hand one ascending to Goatfell showing particularly fine examples of granite jointing. It has been argued that dilatation jointing plays an important role in controlling the evolution of glacial troughs.[11] As glacial erosion processes remove rock from the floor and sides, pressure on the underlying bedrock is reduced. Dilatation joints are a response to

this, developing parallel to the trough walls and floor. Such gently concave joint planes can be found in Glen Rosa, curving up in conformity with the valley sides and intersecting in the arêtes with other sets of joints developed in relation to neighbouring valleys and cirques. Although areas of scree and other superficial deposits can be seen on the photograph which to some extent mask the true bedrock profile of the trough, bedrock is probably nowhere far below the ground surface.

In the late-glacial Loch Lomond Stadial, small glaciers re-developed in the Goatfell group, the one in Glen Rosa extending down-valley almost to the sea, beyond the lower edge of the photograph.[12] Small fresh-looking moraines to be found near the head of Glen Rosa, roughly where the cloud shadow on the left reaches the valley floor, may have formed during a retreat phase of the Loch Lomond ice.

The Tal-y-llyn valley (photograph 26) in North Wales contains a variety of physical features that excite speculation about their evolution, and provide some

26 The Tal-y-llyn valley, Wales. The straightness of this glaciated trough is controlled by a major fault. The lake of Tal-y-llyn is retained by landslide debris which, post-glacially, fell into the valley from the huge arcuate scar on the left-hand valley wall (middle distance, partly in shadow). Also clearly visible is the alluviation at the head of the lake in the foreground, and a small delta entering on the right. SH 711095.

fascinating insights into the events of preglacial, glacial and late-glacial times. The view along this remarkably straight valley, south-westward from the lake of Tal-y-llyn in the foreground to the coast in the distance, stretches for nearly 20 km. Geologically the alignment of the valley is controlled by a major fault, the Bala Fault that can be traced from the Cheshire border across the whole of North Wales to the sea at Towyn. It is presumed that the valley existed preglacially, cut by the Dysynni river along the belt of fault-broken rock. The extent of subsequent glacial modification cannot be determined, but it is likely that ice flowing along it to the south-west in successive glaciations was able to deepen the valley, create some basins along it now concealed by sedimentary infilling, and to steepen the valley sides.

To the right in the distance can be seen the alluvial floor of a parallel valley which carries the lower course of the Dysynni river. After leaving the lake, the Dysynni follows the Tal-y-llyn valley only as far as Abergynolwyn (see below) where it turns abruptly to the right, cutting through the line of hills that otherwise form the right-hand interfluve of the Tal-y-llyn valley to enter on a lower course in the next parallel valley, also fault-controlled. Below Abergynolwyn, the Tal-y-llyn valley is occupied by a smaller stream, the Afon Fathew. This curious drainage pattern is not a simple example of river capture, for there is no reason why, at a level of only 30 m above sea level, the lower Dysynni should attempt to extend its catchment by cutting a valley through a substantial chain of hills without any geological advantage over its neighbour in the Tal-y-llyn valley. The river divide at Abergynolwyn consists of a rock-bar capped by moraine. The diagram below shows some elements of former drainage directions and former watersheds.[13]

Until 1962, the lake of Tal-y-llyn was regarded as the most southerly example of a rock-basin lake in Wales, the rock-bar at its outlet being capped in addition by supposed moraine. At its outlet from the lake, the Dysynni flows in a narrow gorge cut through what appears at first sight to be bedrock. Detailed mapping of the rock structure has shown this view to be incorrect. The barrier consists neither of bedrock nor of moraine, but of a huge mass of fractured and disarranged blocks that represent landslide debris (see p. 63). The source of this great landslide

left The evolution of the river Dysynni and the Tal-y-llyn valley (after Watson).
1. Trough sides.
2. Wind gaps.
3. Former directions of stream flow.
4. Former watershed between the Dysynni and Dyfi basins.

right The Tal-y-llyn bar.
1. Main breaks of slope.
2. Scree-covered slopes.
3. Landslip terraces.
4. Form lines on the bar.
5. Alluvial deposits.
Contours in the trough only, at 30 m intervals.

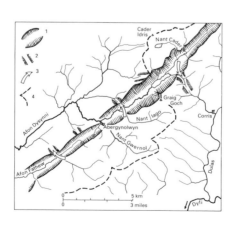

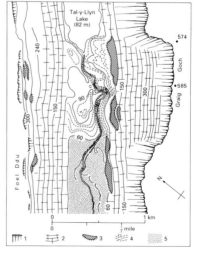

is an arcuate scar on the valley slope, clearly visible on the photograph as a recess above and to the left of the lower end of the lake. The age of the landslide cannot be determined precisely but is likely to be late-glacial, after the Devensian valley glacier had disappeared, leaving the valley sides, over-steepened by glacial erosion, in an unsupported condition prone to collapse. The lake is thus impounded by landslide debris; it does not occupy a basin of glacial erosion though one or more shallow basins may well exist beneath the alluvium of the Dysynni and Fathew valleys. The lake is being steadily infilled by alluvium at its head (foreground) and by an alluvial fan at the right.

A feature commonly described in geomorphological literature as a trough-end is illustrated in photograph 27. The view is looking north-east into the double-headed An Garbh Coire (for location see the map on p. 66). This is an area of the Cairngorm Mountains where deep troughs have been cut by glacial erosion into a relatively smooth upland plateau surface, seen in the foreground and the left of the photograph. The upland plateau here ranges between 1,100 and 1,300 m in height, while the head of An Garbh Coire lies about 300 m below this. Sheet

27 Two overlapping views of the double-headed An Garbh Coire in the Cairngorm Mountains, Scotland. In the foreground and to the left is the upland plateau surface into which the trough-end has been cut. Crossing the photograph in the distance, from left to right, is the glacially eroded trough of Glen Dee, heading (on the extreme left) in the pass of Lairig Ghru (see diagram on p. 66). NN 937982.

Preglacial landforms (shown
by form-lines at 75 m intervals)
and glacial erosion features
(shown by hachures) in the
Cairngorm Mountains. The
approximate extent of the
views shown in photographs
27, 30 and 31 is marked.

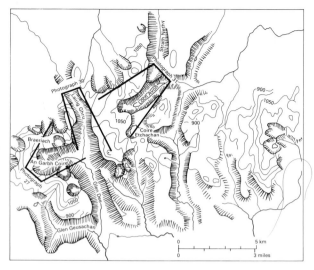

jointing in the granite of which this area is composed lies generally parallel to
the plateau surface, suggesting that the latter is older than the troughs and enabling its general form to be reconstructed. Together with the existence of some
tors and deeply-weathered granite in places, it suggests that the plateau is a preglacial surface of denudation. Strikingly discordant with its gentle contours are the
steep-sided glacial troughs with their ice-smoothed walls. D. E. Sugden has assem-
bled evidence of the patterns of ice flow from the transport of erratics, ice-
moulded bedrock forms, breached watersheds and trough alignments, and has
concluded that the glaciation of the area was of the ice-cap type, the plateau
being almost completely covered at the glacial maximum and the troughs
representing the major routes taken by ice flowing off the plateau.[14] Ice thus
flowed into the head of An Garbh Coire at the glacial maximum; it did not then
originate here, as in the case of a cirque. On the plateau the ice was probably
slow-moving, even immobile in places, as shown by the survival with little modification of preglacial tors, but in the troughs, the ice was streaming rapidly down
selected routeways with their steeper gradients. Doubtless these routeways were
partially prepared for the ice as preglacial valleys, but almost nothing can be
inferred about their initial form. The contrast in behaviour between the ice on
the plateau and the ice in the troughs is remarkable, but such a disparity can
be readily matched today in Greenland and Antarctica where ice moves in streams
within an ice-sheet and can be highly selective in its erosive action.

At a later stage, restricted glaciers occupied some of the Cairngorm cirques;
for example, small morainic ridges lie on the floor of the cirque nearest the lower
edge of the photograph.[15] Their age is uncertain, but probably younger than the
Loch Lomond stage when the Cairngorms were extensively covered in ice. In
the central cirque of the photograph, glacifluvial features are just visible on the
cirque floor, possibly related to decay of the Loch Lomond Stadial ice.

The mountains of southern Scotland (the Southern Uplands) are not so high
as those of the Highlands or north Wales; nevertheless, some of the higher groups
were able to support small independent centres of ice accumulation and dispersal
in the Pleistocene glacials. Systems of cirques and glacial troughs can readily be
recognised, though the dominant lithologies – shales, mudstones and flags – are

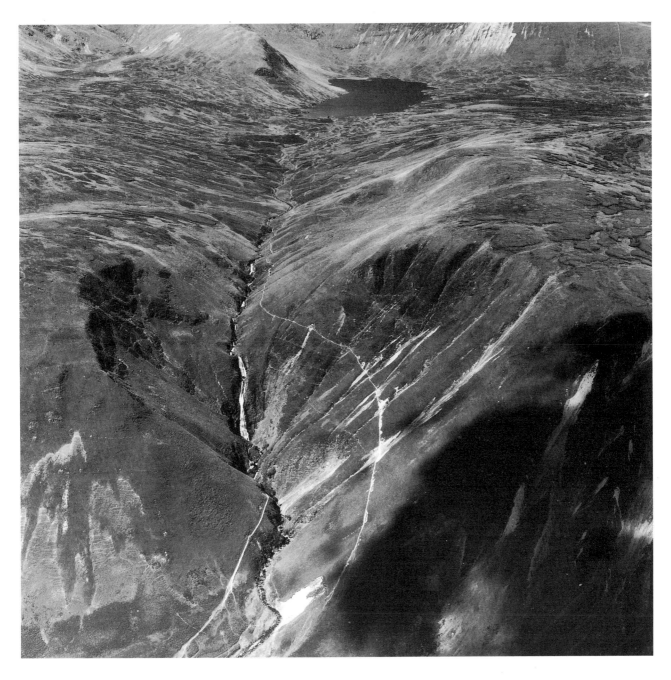

not capable of supporting the spectacular forms presented on some previous photographs. Photograph 28 shows an example of a hanging valley, at whose mouth can be seen the waterfall known as the Grey Mare's Tail in Dumfriesshire. Above the waterfall, the valley of the Tail Burn is broad and climbs only gently from 460 m to the small Loch Skeen at 514 m which occupies a shallow basin at the valley head, surrounded by cirque walls rising to about 750 m. The highest point of this central area of the Southern Uplands is White Coomb, 822 m just beyond the left-hand edge of the photograph. The valley of the Tail Burn hangs about 230 m above the floor of the main valley, the Moffat Water (foreground) to which it is tributary. The Moffat Water lies in a remarkably straight trench,

28 An example of a hanging valley in the Southern Uplands of Scotland. The Tail Burn descends abruptly, by the waterfall known as the Grey Mare's Tail, to the main valley of the Moffat Water in the foreground. The Tail Burn drains from the shallow Loch Skeen, visible in the distance. NT 182150.

running north-east to south-west and draining to the south-west, eroded along a major fault and shatter-belt. The trend is typically Caledonoid, responsible for the main lineaments of the whole of the Southern Uplands. The rocks of this area, whose relatively uniform shale lithology has already been noted, are mainly Silurian, intensely compressed and folded along Caledonoid axes, the dips frequently exceeding 45° and sometimes approaching the vertical. The fissility of the shales and the high angles of dip combine to promote slope instability wherever valley-sides have been sufficiently steepened by fluvial or glacial incision, as can be readily seen in the large landslide scars on either side of the Grey Mare's Tail.

Hanging valleys have long been recognised as typical features of alpine glacial relief.[16] One explanation was that the greater bulk of the trunk glacier was able to cut a larger valley cross-section than those of smaller tributary glaciers, in consequence of which the floor of the main valley was eroded to a lower level leaving the floors of the tributaries 'hanging'. In reality, convincing explanations may not be so simple. Geological differences may have to be taken into account, for example, that the Moffat Water in this case has developed along a weaker shatter-belt whereas the Tail Burn is crossing a series of folds at right-angles to the strike. Another doubt in this area is how far the landforms are solely the product of glacial erosion, or whether the landscape can alternatively be interpreted largely as the work of fluvial denudation slightly modified by glaciation. In the latter case, the 'hanging' elements could be attributed to the arrested migration of a fluvial knick-point, resulting from rapid incision of the Moffat Water along its zone of weakness during tectonic uplift, for instance. Arguments about the evolution of this landscape must also consider the role of multiple glaciation and the long intervening episodes of interglacial fluvial erosion, and must also raise questions about the preglacial landform inheritance. In this area, the upland surface above about 500 m shows a much lower amplitude of relief than the terrain at lower levels, and it may be that it includes elements of one or more ancient peneplains. Entering into the discussion is one other geological element: crossing the photograph from left to right, just down-valley from Loch Skeen, is a dolerite dyke of early Tertiary age. It makes no visible contribution to the relief but its existence shows that the upland surface must be younger than its intrusion, a product of later Tertiary denudation. A possible sequence of events then becomes:

1	Early Tertiary:	igneous intrusion
2	Mid or late Tertiary:	denudation to a surface of low relief
3	Late Tertiary:	uplift and rapid deepening of the Moffat Water valley
4	Pleistocene:	alternating phases of glacial and fluvial erosion; glacial modification of the landscape accentuating and exploiting pre-existing forms
5	Late-glacial:	Deglaciation, slope instability and mass movement

Watershed breaching by ice was described (though not so termed) in north Wales in 1909[17] when it was suggested that the col at the head of the Nantlle valley has been lowered by glacial erosion. Photograph 29 is a view of this col from the west, with the Snowdon massif framing the skyline behind and the upper part of the Nantlle valley (the river is the Afon Drws-y-Coed) in the foreground. In this case there has been no significant change of drainage pattern, since glacial erosion was not sufficiently prolonged to cause total destruction of the watershed: only partial watershed breaching has occurred. The original line of high ground separating Snowdon from the Nantlle valley can be traced clearly in profile on the photograph. From the left-hand edge, a spur of Mynydd Mawr descends at about 8° until it is abruptly broken by the crags of Craig-y-Bera (crest 620 m) which fall with a concave slope to the floor of the col at about 240–250 m. On

29 The glacially breached watershed at the head of the Nantlle valley, north Wales. The view is looking due east towards Snowdon on the skyline; in the right foreground is the shallow Llyn Nantlle-uchaf. The breach has been torn by ice from Snowdon, creating the pass between Mynydd Mawr (left) and Y Garn (the sharp peak on the right). SH 519530.

the right of the col, there is another roughly concave slope steepening to the summit of Y Garn (633 m), beyond which a ridge (partly in shadow) continues to the right-hand edge of the photograph. The form of the relief to the left of Craig-y-Bera and to the right of Y Garn can be used to reconstruct approximately the shape of the former divide, giving a height of 530–540 m for the preglacial col. In successive glacial periods, ice accumulating in the broad basin to the east, fed from the Snowdon massif, overtopped this col and by erosion lowered its level by some 300 m. The rocks involved in this glacial excavation are slates and igneous rocks. The broad basin between Snowdon and the col is probably of preglacial origin, draining with a low gradient to the south, but the escape of the Snowdonian ice in this direction was impeded by the confluence of other ice streams. The form of the basin has suggested an origin under Tertiary conditions of warmer climate and deep weathering, though Pleistocene glaciation has removed the evidence in this locality.

An interesting subsidiary feature of the Nantlle col is the existence of strongly ice-scoured rocky bosses and small basins (one containing a partly artificial lake at 238 m just visible on the photograph). Such a position for a lake is clearly anomalous, but similar features are known from other glacially-eroded cols.[18] The rock bosses rise 50–100 m above the floor of the col. It is possible that severe glacial erosion here was caused by the constriction of the ice within the valley between Craig-y-Bera and Y Garn, and by acceleration of the ice flow as it poured westward into the Nantlle valley.

The concept of watershed breaching by ice, introduced on the previous photograph, is most impressively illustrated in the Scottish Highlands, where many instances have been demonstrated.[19] On a 'provisional sketch-map of the principal glacially-breached watersheds' in the Ochils, Sidlaws, Grampians and part of the Western Highlands, more than one hundred cases were identified, many of which are related to the fact that, at the maximum of the Devensian glaciation, the highest levels of the Scottish ice-cap lay to the east of the main preglacial Highland watershed. Ice was then able to move across the latter to reach the Atlantic, and in doing so, tore out huge breaches in that watershed.

D. L. Linton introduced into the literature of British geomorphology the terms 'glacial diffluence' and 'glacial transfluence', borrowed from the original ideas of the great German/Viennese geomorphologist Albrecht Penck[20] and later developed by Johann Sölch[21] in the eastern Alps. Transfluence refers to the overriding of former watersheds by ice, cutting new escape routes; diffluence refers to the branching of a glacier, whereby part of the ice utilises a lateral col or saddle while the trunk glacier pursues a normal course along the main trough. Photographs 30 and 31 illustrate these ideas respectively from the area of the Cairngorm Mountains. The two sites may be located on the map on p. 66.

Photograph 30 is of the impressive pass known as the Lairig Ghru. Ice pouring off the Braeriach plateau into the two-headed Garbh Coire (photograph 27), together with ice from other cirques and plateau sources, fed a major ice-stream escaping into Glen Dee. The latter, however, was constricted not many kilometres lower down by another powerful glacier issuing from Glen Geusachan, causing the level of the ice in upper Glen Dee to rise. On the northern interfluve of Glen Dee there is evidence of the existence preglacially of a saddle at about 1,070 m.

30 Lairig Ghru, Cairngorm Mountains, Scotland, the finest example in Britain of a new valley created by glacial erosion. On the far left of the photograph in the distance is the summit of Ben Macdhui, the highest point in the Cairngorms (1,309 m). NH 967025.

The ice overtopped this and the upper layers of the Dee glacier began to flow north, eroding the saddle ever more deeply as the thickness of the transfluent ice and its speed increased. Gradually the hugh trough of Lairig Ghru was created, its floor level now reduced to 840 m (at the bend of the valley in the middle distance). This calls for the removal of no less than 230 m of granite. The form of the pass in cross-profile is beautifully parabolic, though the presence of screes is largely responsible for this, and the sharply discordant upper trough edges can be partly seen on the left of the photograph. Notice also the numerous tongues of debris flows and slope failures on the glacially oversteepened slopes. Ben Macdhui (1,309 m), the highest point in the Cairngorms, can be seen at the top left of the photograph.

Photograph 31 shows a classic case of glacial diffluence. The view is taken looking to the south-west, up the valley of Glen Avon, with Ben Macdhui on the left in the far distance. Ice accumulation on the upland plateau in this area discharged into the 200 m deep trough containing Loch Avon, flowing towards the observer. A short distance down-valley from Loch Avon, however, the glacier

31 Loch Avon, Cairngorm Mountains, looking towards the south-west (see the diagram on p. 66). In the lower right of the photograph can be seen the diverging valley of the Saddle, leading into Strath Nethy. Pleistocene ice flowed towards the observer along the trough of Glen Avon; the upper layers of this glacier bifurcated to flow both over the Saddle and along the main trough – a classic example of glacial diffluence. NJ 026032.

was impeded by a competing ice tongue fed from Coire Etchachan. The thickness of the Glen Avon trunk glacier increased as a result, until its surface level exceeded the height of a low pass known as The Saddle, seen in the lower right part of the photograph. Linton reconstructs the preglacial level of the pass as 900 m; its present level is about 810 m, 80 m above the level of Loch Avon. The diffluent ice-stream, departing from the trunk glacier to flow through the Saddle, thus lowered the latter by about 90 m, facilitating its escape into Strath Nethy to the north. The photograph shows clearly how the right-hand wall of the Glen Avon trough is unbroken below the level of the Saddle, continuing beneath the entrance to the Saddle 'as though it were not there'.[22] In contrast, the trough wall at a higher level swings into the Saddle, showing strong ice scouring. There is no doubt, assembling such evidence, that the Saddle is no hanging valley but a clear example of erosion by a diffluent ice-stream.

In areas of low relief where ice flow was mostly unconfined, the impact of glacial erosion can still be distinctive. Huge areas of the Baltic Shield and the Canadian Shield were scoured by the Fennoscandian and Laurentian ice sheets respectively, producing a type of landscape that may also be found, on a more limited scale, in north-west Scotland. Photograph 32 shows part of this landscape around Loch Laxford in Sutherland. The area consists of Pre-Cambrian Lewisian Gneiss, a granitoid gneiss cut by veins and sills of pegmatite and foliated granite. The average dip in this area is steeply to the south-west, Loch Laxford lying roughly along the strike. The photograph shows the strong rectilinear pattern of faults, joints, foliation planes and minor dyke intrusions that has been picked out by erosion. Presumably these lines of weakness were already exploited by shallow preglacial valley systems before glaciation completed the work of differential erosion. The relief of the area ranges from sea level to about 125 m. This low amplitude, typical of most of the Lewisian Gneiss outcrop, is due in large measure to the stripping of a former sedimentary cover, the Torridonian Sandstone (upper Pre-Cambrian), and the exhumation of the sub-Torridonian unconformity. Torridonian outliers survive within 3 km of Loch Laxford. The upper parts of the present land surface fall, for the most part, only a few tens of metres below the level of the unconformity, suggesting that glacial erosion, while it has scoured the surface and produced many distinctive micro- and mesoscale forms, has not in fact lowered the surface dramatically, except for the deepening of some valley lines. The mesoscale forms generally visible on the photograph include scoured rocky bosses and roches moutonnées (sometimes called 'stoss-and-lee' topography),[23] small basins with peat bogs or lakes, fissure valleys and feeble

32 The coast around Loch Laxford, Sutherland, showing a glacially scoured and partly drowned landscape. Strong structural control of the relief is evident, caused by the rectilinear pattern of faults, joints, foliation planes and dykes intruded into the Pre-Cambrian gneiss (Eilean Ard Island: summit at NC 186503, at right centre of photograph).

or non-integrated drainage systems. Locally, slopes can be quite steep, up to 15–20°, while patches of till provide partial infillings of fissures and basins. A general term suggested for such topography is knock-and-lochan relief.[24]

The sea inlets, known as lochs, are analogous to the fjärds of southern Sweden. Basically they represent the drowning during the post-glacial rise in sea level of the outer margins of this glacially-scoured relief, sharply emphasising the strong structural control of relief patterns.

The Aberglaslyn Pass in southern Snowdonia (photograph 33) is the last of a series of rocky defiles along the course of the Glaslyn river before it enters the alluvial plain of Traeth Mawr (foreground). Upstream from the Aberglaslyn gorge lies a small alluvial-floored basin, at the northern edge of which stands the village of Beddgelert, at the point where the Glaslyn is joined by the broad valley of the Nant Colwyn from the north-west. The floor of the Beddgelert basin is almost level at 36–38 m and probably contained a lake in late-glacial times, now silted up. During glacial periods, ice streams from the Colwyn and Glaslyn valleys converged on the basin, the combined glacier being then forced to escape south past Aberglaslyn to Cardigan Bay. Yet at this point the valley narrows so that at the 200 m level it is only 400 m wide. The main cause of this constriction is geological, the river here crossing the outcrop of the Ordovician Lower Rhyolitic Series. The cross-section of the valley from the river up to 200 m is approximately 0.04 km²; that figure should be compared with a mean value for the Glaslyn valley above Beddgelert of 0.19 km² and for the Colwyn valley of 0.26 km², a total of 0.45 km². At the Aberglaslyn Pass, therefore, the valley below the 200 m level is reduced to about one-tenth of its capacity upstream. A compensating increase of ten times in ice velocity is physically impossible, and we must conclude that the greater part of the ice must have been discharged at a level higher than 200 m. The sides of the Pass rise steeply to 200–300 m on the east (right of the photograph) and to about 180 m on the west, at which levels broad shoulders flatten out before rising farther. Most of the Glaslyn-Colwyn ice, then, was discharged at this upper level, that of a broad valley-floor that may have been inherited even from preglacial times though much altered and lowered by glacial erosion since. The steep-sided Pass below this level shows features, such as its V-shaped cross-profile and winding course, that are more typical of fluvial than glacial erosion, and there is no doubt that a considerable part of this fluvial erosion must be attributed to the river in interglacial phases. In glacial episodes when the gorge was occupied by slow-moving ice, subglacially-flowing meltwater must also have played an important part in the further deepening of the valley. There is no evidence that the Beddgelert basin ever contained a lake at a level higher than its present floor, so that a hypothesis of lake overflow to cut the Aberglaslyn Pass must be dismissed.

South of the Pass, the alluvial lowland of Traeth Mawr represents the silting of a former estuary. In the post-glacial period, the sea must have penetrated to the southern exit of Aberglaslyn, for here the level is only 2–3 m above mean tide level today. Silting of the narrow estuary was rapid, however, and especially so after 1811 when an embankment was built at Portmadoc, a few kilometres to the south, shutting out the sea and permitting much of the area to be reclaimed for (rather poor) pasture.

33 *opposite* A view of the Aberglaslyn Pass, north Wales from the south. In the foreground is the beginning of the Holocene alluvium of Traeth Mawr; in the distance beyond the Aberglaslyn Pass is the small Beddgelert basin which in the post-glacial period contained a lake. Glaciers converging on the Beddgelert basin from the left and right distance of the photograph combined to form a massive ice stream that flowed towards the observer, filling the Pass and overtopping the high-level shoulders on either side. Part of the cutting of the Pass is likely to have been the work of subglacial meltwater erosion. SH 594462.

Fluvial as well as glacial processes have contributed to the formation of several of the features already described in this chapter, for instance, in the development of glacial troughs (photograph 26), and in the work of subglacial meltwater cutting the gorge displayed in photograph 33. During the dissolution of the Pleistocene ice in Britain, huge quantities of meltwater were liberated, seeking various routes to escape to ice-free ground. Occasionally, glacial lakes were impounded along the borders of the decaying ice, the water from which sometimes drained subglacially or englacially, and sometimes by utilising an existing col into another drainage system. Photograph 34 shows a classic example of a trench-like valley, Newtondale, cut by meltwater from ice blockading the northern slopes of the North York Moors. The view is towards the south-south-west from near the beginning of the channel (see p. 77); the channel floor in the foreground stands at 166 m, its broad alluvial floor flanked by steep bedrock sides. No through-drainage now utilises the channel, though a minor stream begins to develop alongside the railway. Newtondale was interpreted as a glacial lake overflow channel in 1902.[25] There is now considerable doubt that this simple interpretation, which Kendall and other workers applied to numerous meltwater channels in Britain, is correct, mainly because convincing independent evidence of the former existence of ice-dammed lakes on the scales postulated is lacking. In the case of Newtondale, it seems likely that flat sheets of melting ice supplied most of the meltwater directly, though there may have been a small proglacial lake. The

34 Newtondale, a classic example of a glacial meltwater channel in north Yorkshire, viewed from the north and north-east (see p. 77). Meltwater escaping from melting ice and possibly also small proglacial lakes in Eskdale flooded across a former watershed, cutting this channel with its broad floor and fine swinging meanders, to escape southwards to the Vale of Pickering. SE 840970.

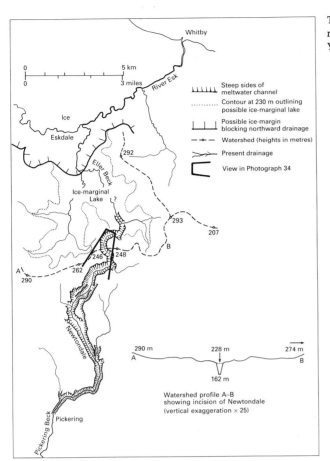

The Newtondale glacial
meltwater channel in the North
York Moors

inset in the diagram above shows how the meltwater utilised an existing col, initially crossing the main watershed of the North York Moors at 228 m. Gradually the channel was deepened by at least 62 m at its head where the water broke out of the Eller Beck valley. It is possible that the channel was initiated subglacially when the North York Moors were still blanketed by decaying ice; as the quantity of meltwater increased and the ice roof collapsed, the channel became occupied by a fast-flowing torrent as broad as the channel itself. The meanders along its course are striking testimony to the power of the stream which finally entered a pre-existing south-flowing drainage system (the Pickering Beck) and thence reached the Vale of Pickering.

GUIDE TO FURTHER READING

D. J. Drewry, *Glacial Geologic Processes*, London, 1986.

C. Embleton and C. A. M. King, *Glacial Geomorphology*, London, 1975.

W. S. B. Paterson, *The Physics of Glaciers*, 2nd edn, 1981.

R. J. Price, 'Glacial erosion', in D. Brunsden and J. C. Doornkamp (eds.), *The Unquiet Landscape*, London, 1972–3.

R. J. Price, *Glacial and Fluvioglacial Landforms*, Edinburgh, 1973.

D. E. Sugden and B. S. John, *Glaciers and Landscape*, London, 1976.

3 Landforms of glacial and fluvioglacial deposition

A corollary of the considerable erosive effect of glacier ice in many parts of Britain is that there must have been an equivalent amount of deposition. Not all of the evidence for this can be found on land, however, and much of the material picked up by the ice lies beyond the present coastline. Nevertheless, about a half of the land area of Britain is covered by glacial and fluvioglacial (glacial meltwater) deposits and, whereas the effects of glacial erosion are most pronounced in highland Britain, the deposits are at their most extensive and thickest in lowland areas. In East Anglia, for example, a thickness of 143 m has been recorded while in the north of the Isle of Man glacial deposits lie 175 m deep. Indeed, parts of east Yorkshire, Lincolnshire and Norfolk owe their very existence to a considerable thickness of glacial deposits. For much of the land surface affected, the form of glacial deposition simply reflects that of the underlying topography but in some areas a distinctive scenery of, for example, drumlins and moraines may be developed. Meltwater streams emerging from the ice margin often deposited great spreads of sand and gravel which may go unnoticed when viewed from the air except, for example, where their thickness is revealed by later downcutting by post-glacial rivers or where they are pockmarked by hollows (kettle holes) denoting the former position of buried ice blocks. Fluvioglacial depositional landforms produced in contact with the ice, on the other hand, may take the form of striking ridges (eskers) and mounds (kames). Whatever the thickness or form of glacial and fluvioglacial deposits they have had a profound influence on aspects of a large part of the British landscape familiar to us today including soils, vegetation, drainage and slope stability.

Until the mid nineteenth century, scientists had not contemplated the former glaciation of Britain, attributing all manner of sediments instead to Noah's Flood. This view changed partly through the observations of British scientists such as Forbes,[1] who visited glaciers in Europe, but in large measure through the persuasive arguments of Louis Agassiz, a Swiss naturalist, who visited Britain in 1840 and recognised glacial effects in upland areas. Acceptance of the 'Glacial Theory' helped to explain the otherwise puzzling transport of boulders both uphill and over great distances. Since they provide such tangible evidence, these 'erratic boulders' (from Latin 'errare', meaning 'to go astray') were vital in reconstructing general ice flow directions and limits. In some cases whole strata have been moved, attesting to the immense transporting power of ice. For example, an entire village appears to have been sited on an enormous erratic of chalk.[2]

Although the timing of glaciations and detailed limits of the ice sheets still remain conjectural, the general pattern of glaciation in Britain had become established by 1894 when James Geikie published the third edition of his book *The*

Great Ice Age and its relation to the antiquity of Man.[3] Despite great quantities of published material, much still remains to be learned about glacial and fluvioglacial deposits, and differences of interpretation remain concerning origins of sediment and even nomenclature. Suffice it here to distinguish broadly between till (formerly 'boulder clay') and fluvioglacial sediments. Till typically comprises pebble- and cobble-sized, frequently angular and sub-angular fragments of different rock types, often in a clayey matrix deposited directly by glacier ice. Fluvioglacial sediments, by contrast, frequently consist of stratified sands and rounded or sub-rounded gravels of varying rock type and are formed through the action of glacial meltwater. The distribution of such sediments in Britain indicates that Pleistocene ice at its maximum extent reached as far south as a meandering line joining London and Bristol and onwards to the north coast of Somerset and Devon and the Isles of Scilly. At the maximum of the last glaciation about 18,000 years ago (see the table below), much of northern and western Britain, East Yorkshire, Lincolnshire and Norfolk was inundated but the Midlands and all of

Chronology of glacial events in Britain[4]

EPOCH	STAGE	EVENT	YEARS BP	CLIMATE	GLACIAL DEVELOPMENTS
HOLOCENE	Flandrian	Postglacial		Temperate (Interglacial)	No glacier ice in Britain
			10,000		
		Loch Lomond Readvance		Glacial/periglacial	Glaciation in upland Britain
			11,000		
		Windermere Interstadial		Temperate	Virtually no glacier ice in Britain[5]
			12,000		
PLEISTOCENE	Devensian Glaciation	Late Devensian Glaciation (Dimlington Stadial)		Glacial/periglacial	Max. extent of Devensian ice sheets, reached about 18,000 years BP
			25,000		
		Early Devensian		Periglacial, but incl. temperate episodes	Cold episode may have given rise to glaciation in highland Britain but no direct evidence[6]
			118,000		
	Ipswichian (last) Interglacial	Ipswichian Interglacial		Temperate	No glacier ice in Britain. Climate slightly warmer than today
			128,000		
	'Wolstonian'[6] Glaciation/Hoxnian Interglacial/Anglian Glaciation/Cromerian Interglacial	Glacial and interglacial episodes		Glacial and Interglacial	Glaciation interrupted by warm periods
	Early Pleistocene	Five glacial periods identified[6]		Alternating glacial and temperate episodes	Four glaciations in W. Midlands and N. Wales and one glaciation in North Sea Region[6]
			2.4 million		

left Glacial limits in Britain. Loch Lomond Readvance (Stadial) limits have been generalised. No glacial limits are shown for Ireland. Locations of photographs are shown.

right Nature of glacial sediments in Britain. Comparison with accompanying map shows the extent to which glacial limits can vary between different authors.

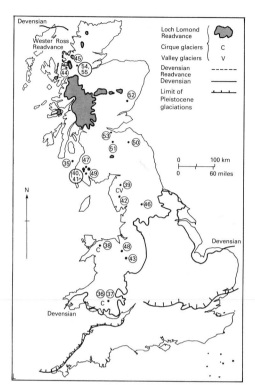

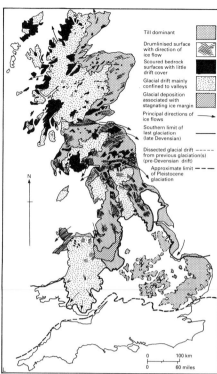

southern England remained unglaciated (see the map above). Additionally, about 11,000 to 10,000 years ago during the so-called Loch Lomond Readvance (Stadial) there was a limited build-up of glaciers once more which was restricted to highland Britain. With greater age, the depositional landforms become progressively less clear in the landscape as a result of weathering and erosion. Landforms from the Loch Lomond Readvance glaciers are the 'freshest' in appearance; those from the last glaciation can still be relatively sharp in outline but those formed earlier in the Pleistocene have been all but obliterated by the long term effects of weathering and erosion; only patchy deposits remain, as indicated on the map above.

Even with the clearest landforms, however, form alone is often a rather unreliable indicator of origin when considering glacial and fluvioglacial depositional landforms. Most of the features depicted in the accompanying photographs are ridges or mounds of one sort or another and yet often very different interpretations are placed on their origins. Compare, for example, photographs 47 and 50; both are sinuous ridges of similar cross-sectional form and size, yet one is interpreted as an esker produced by subglacial meltwater flow and the other as a morainic ridge formed through accumulation of debris at an ice sheet terminus. How then is such a distinction made? The answer lies in taking account not just of form but also of such other factors as the position of the feature with respect to the surrounding topography and to other glacial landforms and sediments in the area, and the stratigraphy and characteristics of the constituent sediments. Yet, frequently the evidence is not unequivocal and opinions may differ quite markedly as to origin.

With these problems in mind, glacial landforms can usefully be subdivided into those aligned parallel to ice flow direction, those aligned transverse to it

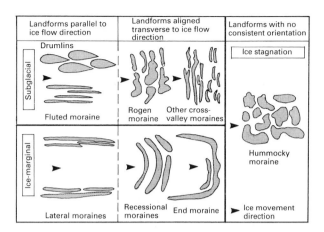

Classification of glacial sediments.

and those with no clear orientation, as shown on the diagram above. A further subdivision is made with respect to conditions of formation, depending on whether the features are produced subglacially by active ice, at the ice margin or through ice stagnation. The term drumlin has been in use in scientific literature for over 100 years and is derived from the Irish Gaelic word 'druim' meaning mound or rounded hill. The ideal three-dimensional shape of a classic drumlin may be likened to that of the inverted bowl of a teaspoon, with the handle pointing up-ice. In reality, drumlins vary considerably in plan form from elongated ridges to almost circular mounds. On average, lengths are less than 1,000 m and heights are generally less than 60 m. They frequently occur in 'swarms' where they may be arranged in ranks or *en échelon* but equally they may join at the base or combine to form double- or triple-crested features. Classically, drumlins are associated with till, but they can also comprise stratified sands and gravels and indeed many would classify similarly-shaped streamlined bedrock forms as 'rock drumlins'.

Rogen moraine, named after Lake Rogen in Sweden, where this feature was first identified,[7] tends to merge with areas of drumlin development. Although aligned transverse to ice flow direction, it is thought to be produced subglacially by active ice. Other types of cross-valley moraine have also been identified.

Fluted moraine is reminiscent of a ploughed field in form, the ridges being aligned parallel to ice flow direction. Ridges found in Britain are often as little as 50 cm high and spaced 1–2 m apart and associated with valley and cirque glaciers of the Loch Lomond Readvance. Lateral moraines, as the name implies, are formed at the sides of glaciers where they form a single ridge, a series of ridges or merely a line of boulders. End moraines are frequently more impressive features than their lateral counterparts and likewise form typically a ridge or series of ridges which in Britain can reach several tens of metres high. They mark the limit of ice extent whereas recessional moraines demarcate halts or minor readvances interrupting a general glacial retreat. Hummocky moraine appears usually as a chaotic arrangement of hummocks and when 'fresh' in appearance is often associated with Loch Lomond Readvance glaciers.[8]

Landforms resulting from meltwater activity can be divided into those produced in contact with ice and those produced by streams flowing beyond the ice front ('proglacial' features). The former are arguably more striking as landforms than the latter and include kame terraces, kames and eskers (see diagram).

right The formation of kame terraces, kames and eskers.

below A proglacial area dominated by fluvioglacial landforms. S – sandur; E – esker; FM – fluted moraine; M – moraine; L – lake.

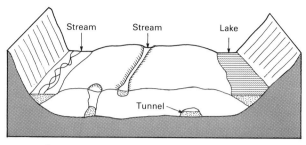

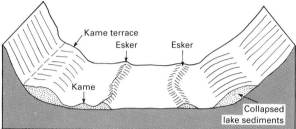

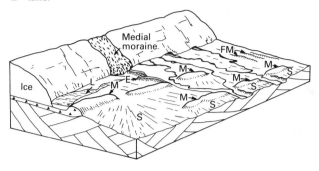

Kame terraces are produced by the accumulation of sands and gravels along the margin of a valley glacier while kames form isolated, often flat-topped mounds resulting from the ingress of sediment-charged meltwater streams into caverns in the stagnating ice. Eskers are sinuous ridges of sand and gravel resulting from meltwater depositing sediment on the ice surface, in a tunnel within the ice mass or at its base. Meltwater issuing from an ice front may spread sand and gravel between confining valley walls leaving a 'valley train', or, where it flows onto a plain, it may form a 'sandur' (Icelandic term for outwash plain; pl. = 'sandar'). A block diagram displaying a range of fluvioglacial landforms is shown above. With the exception of the first photograph, the remainder are dealt with in the following manner. First, a selection of depositional features associated with the Loch Lomond Readvance is considered. This was the last period when glaciers affected Britain, and consequently the resulting depositional landforms tend to be relatively compact and fresh and therefore most readily appreciated. Anyone familiar with moraines found in front of modern glaciers cannot but be struck by the well-preserved nature of many such features: stripped of their vegetation cover, one could easily be misled into believing that the glaciers had only just retreated from the area. Little wonder then that it was features associated with Loch Lomond Readvance glaciers that convinced Louis Agassiz last century that Britain had once nurtured glacier ice. Features from the Brecon Beacons, Snowdonia, Lake District and the Southern Uplands are used to illustrate the range of landforms produced during this period. Second, moraines formed during proposed readvances of the last (Late Devensian) ice sheets are dealt with. Third, landforms produced directly by active ice sheets are considered followed by, fourth, fluvioglacial landforms resulting from the decay of the ice sheets.

Photograph 35 shows the small island of Ailsa Craig off the Ayrshire coast which, although it consists almost entirely of rock, is nevertheless renowned amongst geomorphologists studying glacial deposits in Britain. It is formed of riebeckite-eurite, which is a distinctive microgranite with small dark bluish crystals set in a greyish background. The rock is unlike virtually any other outcrop in the British Isles so that any fragments found in glacial deposits have almost

certainly emanated from this small island. Erratics of the microgranite were first identified on the Isle of Man late last century, and then subsequently in Dublin Bay and all around the Irish Sea as far as Pembrokeshire and Tramore Bay (see the map on p. 84). They have also been found on south Kintyre. Both frequency and size of the erratics tend to diminish with distance from the source so that cobble and boulder erratics can be found relatively easily in Northern Ireland and the Isle of Man: a concentrated search in South Wales, however, reveals only the occasional pebble-sized specimen. Clearly the general direction of transport has been southwards but some erratics have also been moved due west and even slightly north of west of the source. This fan-like distribution is a common feature of many erratic trains although that of Ailsa Craig microgranite is particularly dispersed. One suggestion made early this century to account for the wide dispersal pattern involved carriage of the erratics by ice floes then redistribution by glacier ice. Another suggestion has been that ice sheet build-up during the last glaciation was not synchronous throughout Britain, and that the main zone of snow-bearing winds and consequent main area of ice sheet growth moved southwards across the country. Conceivably, therefore, Ailsa Craig erratics could have been transferred from one area of local ice dominance to the next, without the continuous flow of ice from the Scottish Highlands through to southern Britain,[9] although sea ice may have redistributed some erratics now found in coastal locations. It should be noted that the map on p. 84 shows Ailsa Craig erratics reaching Gower[10] and North Devon,[11] which represents an extension of

35 Ailsa Craig with the Isle of Arran in the background. An island, 11 km off the Ayrshire Coast, it consists of a distinctive microgranitic rock type known as riebeckite-eurite for which there is no other known outcrop in the British Isles. Glacial erratics from this island have been found all around the Irish Sea Basin, and beyond to Bristol Channel coasts (see the map on p. 84). NX 018998, looking N.

Distribution of Ailsa Craig erratics.

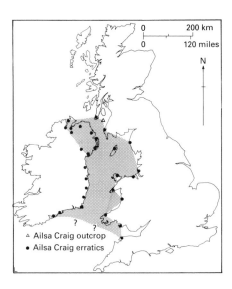

the distribution compared with earlier maps. It may well be that the actual distribution is even more extensive but establishing this depends on chance finds by geomorphologists or geologists familiar with the rock type.

LOCH LOMOND READVANCE LANDFORMS

The sequence of morainic ridges impounding Llyn-cwm-llwch in the Brecon Beacons (photograph 36) provides a good illustration of the remarkable freshness of Loch Lomond Readvance glacial depositional features in upland Britain. Cwm Llwch is just one of a number of spectacular glacial cirques along the north face of the Old Red Sandstone escarpment of the Brecon Beacons. W. S. Symonds in 1872 first drew attention to the Cwm Llwch moraines,[12] and later T. Mellard Reade (1894–5) wrote that 'the sun has traced out this moraine, and settled its alignment and position in the larger Cwm or Valley'.[13] Certainly shading is one aspect explaining the huddling of the glacier in the extreme south-western corner of the escarpment face but the extensive upland slope to the south-west, from which snow could have drifted into the cwm, is another factor. In addition, the curvature of the headwall in the foreground of the photograph would have tended to protect snow in its lee from the scouring action of the wind.

Lewis considered that there were two sequences of moraines in Cwm Llwch; a dubious set of two very subdued features extending along the base of the escarpment towards Corn Dû and a clear sequence of five fresh-looking moraine ridges grouped around the lake.[14] The latter sequence rises to more than 46 m above the valley floor along its western limb and at its greatest elevation it lies at about 580 m. It covers about 395 m in an east–west direction and the outer ridge is only 366 m from the backwall. The angle of elevation from the top of the outermost point of the moraines to the top of the backwall along the glacier axis is 16° which, Lewis argued, suggested the presence of a steep glacier that melted vigorously at its snout and as a result perhaps moved rapidly throughout its length. Such a regime would certainly be typical of a maritime glacier with a high precipitation input counterbalanced by considerable summer ablation. These

36 *opposite* Cwm Llwch, Brecon Beacons, South Wales. The sequence of arcuate, fresh-looking cirque moraines damming the lake, Llyn-cwm-llwch, can be seen. These moraines are thought to have been formed by a cirque glacier that occupied part of the escarpment about 10–11,000 years ago during the Loch Lomond Readvance (Stadial). The escarpment rises towards Corn Dû (873 m) in the upper left of the photograph, which is the second highest point in the Brecon Beacons. The near-horizontal bedding of the Old Red Sandstone bedrock is clearly shown on the hillslopes in the lower right of the photograph. Cwm Llwch at SO 002220, looking SE.

fresh-looking moraines were probably formed by a glacier that occupied the cirque during the Loch Lomond Readvance or Stadial (11–10,000 years BP).

Although the valley head of Cwm Crew, located about 2 km south of Cwm Llwch in the Brecon Beacons, is at a greater elevation than Cwm Llwch (*c.* 640– 762 m as opposed to *c.* 580 m), one might be forgiven in a cursory perusal of a large-scale map for rejecting this site as a likely location for a glacier during Loch Lomond Readvance times. The valley head faces almost directly south so that there is little protection from the sun, whereas the Cwm Llwch glacier would have remained in the shade virtually all day. The shape of the valley head does not conform to the classic cirque form (cf. photographs 38 and 36): it is elongated down-valley, the only steep crags present are at most a mere 50 m high and in any case they are located to one side of the valley head and there is no tendency

37 Cwm Crew, Brecon Beacons, South Wales. The steep-sided outer moraine from a glacier of Loch Lomond Readvance age can be traced down the left-hand side of the photograph towards the Nant Crew stream. It can be seen on the east side of the stream gradually diminishing in height as it runs away from the camera position. Cwm Crew at SO 008198, looking NW.

for a flattening of the floor. On the ground or from the air (photograph 37), however, the impressive fresh-looking morainic sequence, and in particular the outer ridge, would lead one to reconsider the initial rejection of this site. Along the eastern side of the valley head, this outer ridge increases in height southwards until it turns westwards to isolate the valley head from the lower portion of the valley. Inside this ridge at its southern end, a further six much smaller ridges have been distinguished, which are now dissected by the Nant Crew stream. The configuration of this moraine allows us to reconstruct the former position of the small glacier that must have occupied this hollow *c.* 11–10,000 years ago. Despite facing south, the valley head has certain advantages with respect to snow accumulation. First, a major conclusion to emerge from the work of J. B. Sissons[15] concerning the palaeoclimatic inferences that could be deduced from the pattern of Loch Lomond Readvance glaciers in Scotland was that snowfall was associated with south to south-easterly airstreams preceding the warm or occluded fronts of depressions in addition to winds from a south-westerly direction. This may help to explain in part the build-up of the Cwm Crew glacier in this apparently unfavourable position. Second, the large, gently-sloping expanse of Cefn Crew immediately to windward of the prevailing south-westerly winds would have given plenty of scope for snow-blowing. Drifting snow would have tended to accumulate on the west side of the valley below the crags, building out eastwards across the valley head, but ice movement would have been down-valley. This direction of ice movement is reflected in the increase in size and lobe-like extension of the morainic ridges down-valley.

Of the many cirques in Snowdonia, North Wales, one of the most frequented for scientific purposes is Cwm Idwal at the southern end of Nant Ffrancon (photograph 38). For many, this cirque ranks as a classic example. It is excavated along a syncline (downfold) of Palaeozoic rocks to a depth of 450 m below the neighbouring Glyder plateau. It has very steep rock walls surrounding an almost level floor nearly 1 km long, much of which is occupied by Llyn Idwal, a shallow lake. During the Loch Lomond Stadial, this cirque was occupied by glacier ice together with more than 30 other cirque glaciers in Snowdonia (see diagram below). Some aspects of Cwm Idwal are atypical of other cirques in the area. It is considerably deeper and larger than the other 14 cirques of the Glyder range in the vicinity; its floor at 360 m, descends some 180–300 m below most of its neighbours and its enclosed area of 1.9 km² exceeds that of the next largest cirque almost two-fold. This abnormal size seems to result from its complex origin. Far from being a simple cirque, the size of Cwm Idwal has been enhanced by ice from the two smaller cirques of Cwm Cneifion and Cwm Clŷd cut high into the flanks of its backwall (see diagram on p. 88). In addition, Cwm Idwal acted as a diffluent trough during the Devensian glaciation, when ice streamed over the backwall via Twll Du contributing both to the erosion of the backwall and to the excavation of the basin.

This complex form may explain the far from easily interpreted arrangement of depositional mounds and ridges seen in and around Cwm Idwal. In his autobiography, Charles Darwin said of a visit he made to Cwm Idwal in 1831 with Adam Sedgwick:

> On this tour I had a striking instance how easy it is to overlook phenomena, however conspicuous, before they have been observed by anyone . . . neither of us saw a trace of the wonderful glacial phenomena all around us; we did not notice the lateral and terminal moraines . . . yet these phenomena are so conspicuous that . . . a house burnt down by fire did not tell its story more plainly than did the valley.[16]

However, more than 150 years later, the story is far from resolved and controversy continues. First, the age of the ridges is not entirely clear, although most experts would agree that the ridges seen in photograph 38 were formed during Loch Lomond Readvance times. Second, the origin of the moraines is also unclear: indeed some have disputed whether all are moraines. Four major groups of ridges can be discerned, some more distinct than others (see the diagram on p. 88). Moraine 1 is a rather subdued feature on the lip of the cirque, its size being exaggerated by the underlying bedrock and its inner (or proximal) slope having been affected by a higher ice-dammed lake level than that of the present day

Llyn Idwal.[17] The subdued nature of this moraine and the adjacent set (1a) has led to speculation of a pre-Loch Lomond Readvance date for their formation, although this has been disputed.[18] The hummocky moraines (2) straddling the narrow part of Llyn Idwal are clearly seen in photograph 38. The more subdued lateral moraines (2a), however, that ascend the eastern cliffs above the main group 2 moraines are arguably more important in unravelling the sequence of formation of the entire pattern of ridges in Cwm Idwal, as will be discussed below. Two small ridges (3) probably represent a retreat phase of a glacier in Cwm Clŷd. Most dispute concerning the origin of the ridges and mounds has focussed on a long parallel ridge sequence (4) above the western shores of Llyn Idwal. Three different origins have been suggested: that they, (1) formed through accumulation of debris at the base of a snowpatch and represent therefore a protalus rampart,[19] (2) are fluted moraine formed at the base of fast-moving ice,[20] or (3) represent lateral moraines. Since the ridges contain erratic breccias and tuffs from the eastern cliffs of Cwm Idwal rather than from the adjacent western cliffs this argues against a protalus origin.[21] There are also problems with the shape of the ridges in plan form if a protalus rampart origin is envisaged. That they represent fluted moraine must also be in doubt for a number of reasons, including their asymmetrical cross-profile with steeper eastern than western slopes and their extremely large size – much larger than is acceptable with current glaciological theory. The ridges do, however, resemble closely lateral moraines formed elsewhere. Asymmetry is typical of lateral moraines, the steeper side being originally supported by the glacier. The main difficulty with this mode of origin is explaining the rather unusual position of the ridges. If one envisages Cwm Idwal occupied by a cirque glacier, then the position of these ridges might seem rather low and they seem to terminate rather inexplicably at the base of the backwall rather than climbing steeply up the western crags. However, if Cwm Cneifion contributed considerable quantities of ice to the Idwal glacier, then at least a component of glacial movement would be diverted towards the western slopes, and the position of the moraines low down on the western slopes is easily comprehended. This view is supported by the erratic content of the ridges.

In all, 64 former glaciers dating from Loch Lomond Readvance times have been identified in the Lake District (see the diagram on p. 90). They range in

left Mapped limits of former Loch Lomond Readvance glaciers in Snowdonia.

right Moraines in Cwm Idwal. Moraine groups 1–4 are shown (see text for explanation). The diagram is orientated to help with interpretation of photograph 38.

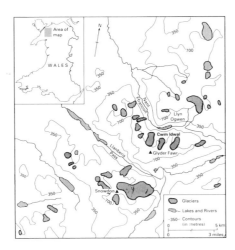

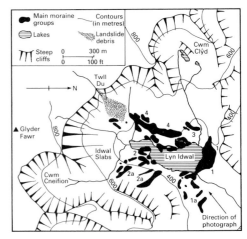

38 Cwm Idwal, Snowdonia, North Wales. The origins of the depositional features associated with this cirque have been the subject of much debate. Most experts agree that they are all moraines of some sort, and that they date from Loch Lomond Readvance times. The long, slightly curving parallel ridges of moraine group 4 lie beyond the shallow lake (Llyn Idwal) on the floor of this glacial cirque. Moraine group 2 straddles Llyn Idwal at the point where the lake narrows. Less obvious but still discernible are some of the other moraine groups identified in the diagram on p. 88. Cwm Clŷd (upper right), the steep backwall of Cwm Idwal cut through a syncline (downfold) of the Palaeozoic rocks and the fossil landslide below Twll Du can be seen just left of the centre. A possible protalus rampart may just be discerned on the slopes above moraine group 4 near the centre of the photograph. Cwm Idwal at SH 645598, looking SW.

size from valley glaciers, such as those in upper Grisedale and Deepdale with areas of 3.18 and 2.43 km² respectively down to small cirque glaciers occupying comparatively small hollows, such as Black Combe (0.08 km²). Another small glacier, and the most northerly of all the mapped glaciers in the Lake District, occupied the Bowscale Tarn cirque. It had an area of only 0.13 km², yet its former terminus is marked by a massive end moraine, some 120 m broad and 15 m high on both sides (photograph 39). The Wolf Crags glacier, some 7 km north of Grisedale (as shown in the diagram on p. 90), is similarly small and it too has a well developed end moraine 40–150 m broad, 2–7 m high and over 1 km long. The apparent paradox of these small glaciers producing large end moraines has been

Mapped limits of former Loch Lomond Readvance glaciers in the Lake District.

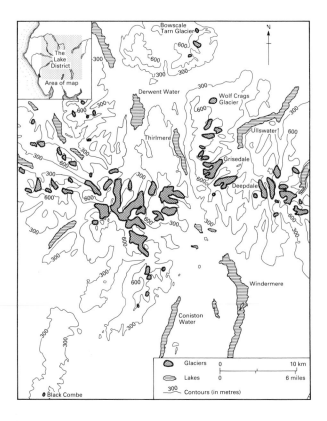

given the following explanation.[22] Such impressive barriers imply that the glaciers, though active, had approximately stable margins for a very long time. These two glaciers were flanked to the south-west and south-east by large expanses of hillside which would have provided sources of nourishment from wind-drifting snow. Once the snow had built up to the top of the headwall, it is argued, further snow would have continued down the smooth glacier surfaces to be distributed over the ground beyond the glacier snouts, thus resulting in approximately stable margins.

In contrast with the considerable development of glacier ice at the time of the Loch Lomond Readvance in the Scottish Highlands, the Southern Uplands saw the build-up of very little glacier ice. An arcuate, double-end moraine marks the limit of a Loch Lomond Readvance glacier that spilled out of a large cirque-like depression between the hillmasses of Mullwharchar and Hooden Hill, some 5 km south of Loch Doon in the Gala Lane Valley, western Southern Uplands (photograph 40). The outer moraine is larger than the previous examples, being 200 m broad at its widest point and its steep outer slope has suffered erosion by the Gala Lane river, which has accentuated the moraine slope as a result. The moraine reaches its maximum height of 10–12 m at the southern end where it comprises ridges and depressions of about 2–3 m difference in height. The moraine is less prominent northwards, where it becomes increasingly covered with angular boulders of granite. Several ridges on the moraine parallel the curving form of the outer moraine and they reflect an oscillating margin to the glacier. This broad outer moraine is paralleled on its inner margin by a delicate, sinuous morainic ridge about 3–5 m in height, and this recessional moraine represents a stand-still or minor advance of the glacier as it retreated from the outer moraine. The east-facing aspect, sheltered from the prevailing westerly winds, and steep nature of the 300 m high backwall of the cirque-like depression would have offered an extremely favourable collecting area for snow blown off the broad expanse of hillslopes lying to the south-west. Indeed, the potential snow-blowing area to the south-west is at least as large as that of the glacier itself, which would explain why this glacier descends to such a low altitude (250 m) when compared with the other Loch Lomond Readvance glaciers in this part of the Southern Uplands (295–550 m). This is shown in the diagram on p. 92.

Another Loch Lomond Readvance glacier moved into the valley of the upper reaches of Elgin Lane, about 2 km south-west of the previous example. This particular glacier, however, left no end moraine sequence to mark its former

39 Bowscale Fell, Lake District. The River Caldew is seen in the foreground with Bowscale Fell forming the high ground at upper left of the photograph. The impressive end moraine formed by a glacier that occupied the cirque during Loch Lomond Readvance times all but obscures Bowscale Tarn (centre). NY 337313, looking SW.

40 Mullwharchar, Ayrshire. The moraine complex of Loch Lomond Readvance age can be seen as a belt of hummocky terrain on the far side of the Gala Lane stream flowing from left to right across the centre of the photograph. NX 467876, looking SW.

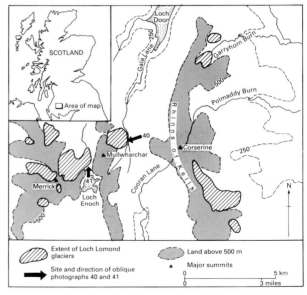

Distribution of former Loch Lomond Readvance glaciers in the western Southern Uplands.

existence, but instead deposited a remarkable sequence of hummocks. These hummocks are typically about 3 m in height and 50 m in length and overall show no consistent orientation. These features are referred to quite simply as hummocky moraine. In photograph 41, the hummocks can be seen to occupy a sweeping zone of low ground from the middle left (mouth of the tributary valley of Caldron Burn) extending northwards down the Elgin Lane Valley where they terminate abruptly some 300 m short of the junction of Elgin Lane with Saugh Burn (upper centre of photograph). In the absence of an end moraine, the southern limit of the glacier, which had its source in two cirque depressions cut in the eastern face of Merrick, has been delimited in the foreground of the photograph by the abrupt edge to the hummocky moraine. In the upper left of the photograph, on the Rig of Munshalloch, however, the glacier extended upslope beyond the hummocky moraine where its limit can be traced by the characteristic deposits that it left behind. The hummocks comprise a friable till with a gritty matrix and they are strewn with boulders of both granite and greywacke. The presence of the latter confirms the contrasting direction of local ice movement during Loch Lomond Readvance times as opposed to full glaciation, for the greywacke boulders could only have been transported from the east and south-east whereas ice-sheet movement during the Devensian glaciation was southwards into the area from an ice divide in the vicinity of Loch Doon.

During the Devensian glaciation, the Loch Doon basin acted as a major centre of ice sheet accumulation and competed vigorously to the north with ice derived

41 Elgin Lane, Kirkcudbrightshire/Ayrshire. Hummocky moraine, deposited by a Loch Lomond Readvance glacier emanating from the Merrick hillmass off to the left of the photograph. NX 445867, looking N.

from the Highlands. In the east, Southern Uplands ice extended nearly to Edinburgh and produced a major assemblage of ice-moulded features in the Tweed basin. Yet, during the Loch Lomond Readvance there were only a few minor glaciers in the Southern Uplands leaving evidence like that featured in the accompanying photographs. This contrasts with the huge ice masses developed at that time in the Scottish Highlands (see the diagram on p. 80). This discrepancy requires an explanation. The only feasible answer seems to rest on there being a marked difference in the behaviour of snow-bearing winds during the last glaciation as compared with Loch Lomond Readvance times. It is surmised that during the (major) Devensian ice sheet build-up, the track of major snow-bearing winds must have crossed, first, the Scottish Highlands, then the Southern Uplands and the Lake District, finally reaching as far south as Wales. The minute amount of ice in the Southern Uplands and other uplands areas to the south during the Loch Lomond Readvance compared with farther north, on the other hand, suggests that at this time the track of major snow-bearing winds never moved southwards beyond the Scottish Highlands. As a result, these other upland areas of Britain, including the Southern Uplands, received considerably less in the way of snowfall.[23]

Just as hummocky moraine can be found inside the limits of many Loch Lomond Readvance glaciers in Britain, so too can fluted moraine. However, whereas hummocky moraine is unlikely to be missed by the casual observer on the ground even if its origin is not appreciated, fluted moraine may well go unnoticed because the individual ridges are small, often no more than 0.5 m high and 2 m apart. Oblique aerial photography might require very special lighting conditions to reveal its pattern but on vertical aerial photographs their pattern can show up quite well. Photograph 42 shows fluted moraine in part of Upper Oxendale in the English Lake District. This fluted moraine was deposited by a glacier of about 3.5 km² that reached some way down Oxendale, its terminus being marked by a series of mounds. A lobe of ice also spilt over southwards through the col between Cold Pike and Pike of Blisco and terminated in Wrynose Pass.[24]

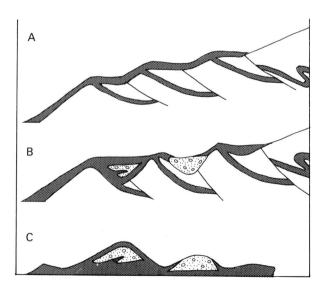

Generalised model showing how the Ellesmere–Whitchurch moraine may have formed. (A) Initial system of ridges and troughs in which outwash and flow till accumulate; (B) after deglaciation; (C) trough fillings become ridges which contain flow till and outwash in various proportions.

42 Upper Oxendale, Lake District. Fluted moraine formed by a Loch Lomond Readvance glacier. The photograph covers an area approximately 480 × 460 m. SD 256055.

Fluted moraine deposited by Loch Lomond Readvance glaciers often occurs on steep or fairly steep, till-covered slopes where the glacier would have been flowing relatively rapidly. Fluted moraine can also be found emerging from beneath present-day retreating glaciers, the ridge alignment also paralleling ice movement direction. These modern examples, however, form on horizontal or gently-sloping surfaces rather than on steep slopes, suggesting that they have formed in a quite different way.

MORAINES FORMED BY PROPOSED LATE DEVENSIAN READVANCES

Whilst the Loch Lomond Readvance is widely accepted and its timing and conditions reasonably well understood, the status of various proposed readvances of the last ice sheets remains a matter of controversy. Names such as the 'Aberdeen–Lammermuir' and 'Perth' readvances will be familiar to anyone with a strong interest over the last 20 years in the glacial history of Britain. Yet, these readvances were shown to be unfounded by the mid-1970s,[25] the latter being based on widespread fluvioglacial deposits which feature in photographs dealt with later in this chapter. Depositional landforms relating to two other proposed readvances are shown in photographs 43–5. The first photograph depicts part of what has been referred to as the Ellesmere Readvance moraine,[26] although more recently[27]

the landforms have been viewed as representing glacial deposition on a downwasting ice margin during the retreat of the last ice sheet from its maximum extent. Photographs 42 and 43 show parts of an extensively developed end moraine along the coasts of Wester Ross, north-west Scotland (see diagram on p. 100). It has recently been proposed that this moraine represents a readvance of the last ice sheet. So far there has been no corroborative stratigraphic evidence, so that it may represent a phase of standstill rather than of readvance.

Photograph 43 shows part of the Ellesmere–Whitchurch moraine complex south-east of Ellesmere itself. Here, the moraine comprises a chaotic landscape of hillocks and hollows, many of which hold meres (lakes). The relief of this part of the moraine is up to about 40 m. The moraine section shown is just a small part of a sweeping arc of such topography that loops southwards from near Wrexham towards Ellesmere and then north-east towards Whitchurch where there is another arcuate band of similar topography which can be traced eastwards to Bar Hill. This twin loop, which is as much as 3 km wide in parts, is thought to represent a phase of retreat of Late Devensian ice. The maximum limit of this ice is thought to be around Church Stretton and Bridgnorth, a further 40 km to the south. Erratics in the glacial sediments indicate that the ice producing this moraine moved southwards from the Irish Sea Basin in the vicinity of the Dee Estuary. However, Welsh erratics amongst the deposits also imply that there was some competition for supremacy between ice masses from Wales and from the Irish Sea Basin in this area. The band of chaotic hillocks and hollows is thought to have resulted from widespread stagnation of the ice margin. Many sections in the moraine, such as those from the Wood Lane sand and gravel quarry, show a complex sequence of sands, gravel and till, which may be interpreted as reflecting a predominantly supraglacial origin,[28] with material brought to the ice surface and then accumulating in troughs in the ice (see the diagram on p. 94). These trough accumulations now form hillocks in the landscape while the subsequent melting out of buried ice masses would have led to the formation of the hollows. Many of the tills seen in the exposures have been interpreted as flow tills which, as the term implies, were formed by sludging or flowing of the sediment over the ice surface. Such sediments are regarded as typical of a stagnating ice margin.

Since the late 1970s, compelling morphological evidence has been published for a readvance of the last ice sheet in Wester Ross, north-west Scotland.[29] The principal evidence of the former ice limit consists of extensive ice marginal moraines in five areas of Wester Ross (see the map on p. 100). These moraines are unlike the complex Ellesmere–Whitchurch moraine, comprising frequently stark, single ridges crossing otherwise almost bare glacially abraded bedrock. That they are regarded as moraines and not, for example, eskers, is indicated by their constituent sediments: they consist of till rather than sands and gravel. It would be reasonable to question further as to whether these ridges are not in fact Loch Lomond Readvance glacier limits rather than Devensian readvance features, given that this area nurtured glacier ice down to low altitudes during the more recent period. Certainly, mapped Loch Lomond Readvance glacier limits do occur in the area but they are associated with localised ice masses whereas the mapped Wester Ross Readvance moraine is in no way related to these limits, lying as it does some way intermediate between the Late Devensian ice sheet and the Loch

43 *opposite* Ellesmere–Whitchurch moraine, Shropshire. White Mere lies at the extreme left of centre of the photograph with Wood Lane sand and gravel pit near the centre. SJ 417323, looking N.

44 Applecross moraine, Wester Ross, north-west Scotland. The Wester Ross Readvance moraine can be seen to the right of centre of the photograph where it forms a single ridge. It curves around to the east-north-east on the lower slopes of the escarpment of Croic-bheinn where it forms an area of boulder-strewn hummocks. Loch Gaineamhach lies left of the centre. NG 764521, looking SE.

Lomond glacier limit. Indeed, one crucial piece of evidence demonstrates the earlier formation and lack of continuity of the Western Ross Readvance moraine with the Loch Lomond Readvance glacier limits featured in photograph 45: this evidence consists of a Loch Lomond glacier which appears actually to have cut through the earlier Readvance moraine.

One section of the Western Ross Readvance moraine occurs on the Applecross Peninsula. Photograph 44 shows part of this Applecross moraine where it follows a shallow concave up-valley curve for *c.* 2 km north-west of Croic-bheinn around Loch Gaineamhach. At the base of the scarp face of Croic-bheinn, the moraine consists of an area of boulder-strewn hummocks bounded on the south side by a ridge. Farther west, the ridge widens into a steep-sided complex feature over 15 m high and 200 m wide with up to five parallel crests. To the north-west (foreground of the photograph), the moraine is restricted to a single well-defined ridge over 10 m high. The moraine is strewn with large boulders up to 4 m in length and comprises sub-angular to rounded boulders and cobbles of the local bedrock (Torridonian Sandstone) together with occasional erratics of Lewisian Gneiss.

On the lower slopes of the north-western end of the Baosbheinn ridge, near Loch Maree, Wester Ross, three depositional ridges can be found (photograph

45). Two of these are interpreted as moraine fragments formed during the Wester Ross Readvance and the larger feature as a protalus rampart formed at the base of a snowbed under periglacial conditions, which is the topic dealt with in the following chapter. These features illustrate well the problem of determining origin on the basis of form. Originally the larger ridges were both accorded a periglacial origin on the basis mainly of form, but more recent research, involving sedimentary analysis of all three ridges, has demonstrated a morainic origin for the lower two ridges, an interpretation which is supported strongly by their content of erratic boulders and fine sediment, the latter being uncharacteristic of protalus ramparts, since fine material cannot slide or bounce down a snowbed. The moraine remnants also show strong similarities in their sedimentary characteristics with other Wester Ross Readvance moraines.[30]

Viewed on a larger scale, these moraine remnants make a logical extension of existing limits defined for the Wester Ross Readvance in the area, except for the question as to why they terminate so abruptly at the south-western end (right side of photograph). The answer lies in the extent of glacier ice during the later Loch Lomond Readvance. Evidence suggests that a large ice field built up over a wide area between Beinn Alligin and Beinn Eighe and two outlet lobes of this

45 Baosbheinn ridge, Wester Ross, north-west Scotland. Below the escarpment of Creag an Fhithich (737 m) in the foreground, can be seen a large boulder strewn ridge regarded as a protalus rampart formed under periglacial conditions. Two more subdued ridges lie below this ridge to the west. These are Wester Ross Readvance moraine fragments truncated by Loch Lomond Readvance glaciers (see diagram on p. 100). NG 854678, looking SE.

Glacial limits in north-west Scotland. Loch Lomond Readvance and Wester Ross Readvance limits near Baosbheinn, Wester Ross (top), and Wester Ross Readvance limits in north-west Scotland (bottom).

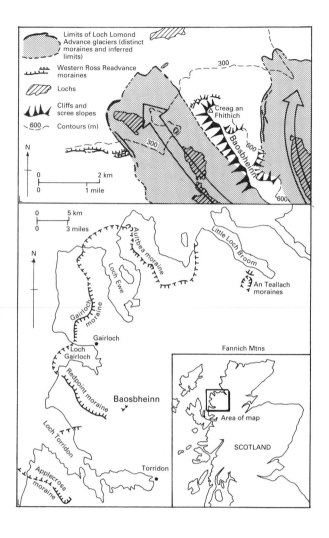

ice field were deflected either side of the Baosbheinn ridge. The lobe on the south side of the ridge cut through the Wester Ross Readvance moraine as it advanced a further 2 km to the north-west (see the map above).

GLACIAL LANDFORMS FORMED BENEATH ACTIVE LATE DEVENSIAN ICE

Of all the features deposited by the last ice sheets, arguably the drumlin would rank as the most evocative of the former passage of an ice sheet. In this respect, drumlins are to glacial deposition what cirques are to glacial erosion: both are unmistakably the work of ice. Seen from the ground, one is aware of the 'whale-back' form of drumlins but to appreciate best their three-dimensional streamlined form, they should be viewed from the air. Photograph 46 shows a well-developed field of drumlins in Ribblehead, Yorkshire. In producing them, an ice stream moved from near Langstrothdale Chase (right of the photograph), at which point it was flowing westwards. As it entered Ribblehead, it swung gradually round to the south-west (at the viewpoint in the photograph) and then to the south to move down Ribblesdale where it was joined by other ice streams emerging from tributary valleys and deflected southwards as they entered Ribblesdale from

the east. These changes in the direction of ice flow are clearly recorded in the orientations of the long axes of the drumlins in the area. Not only are drumlins elongated and therefore streamlined in the direction of ice movement but they also classically display another aspect in common with features offering least resistance to flow in moving fluids: they usually have a 'snub' end facing the direction of flow and a tapering one facing downstream. Analogies with this aspect of drumlin form can be found in, for example, the submarine or aircraft wing, both with a 'snub' end facing into the moving fluid. The north to south movement of ice in Ribblesdale, so graphically illustrated by the alignment of the drumlins, is also indicated by the direction of striae on bedrock exposures and by the direction of transport of erratics in the area, including the famous Norber erratics, as shown in the following diagram. The curved, sub-parallel tributary valleys of Ribblesdale, of which Langstrothdale is one, have no sign of clear structural control and therefore suggest active and highly effective glacial erosion. That the drumlins, the epitome of active glacial deposition, occur in close proximity

46 Drumlins, Ribblehead, Yorkshire. Langstrothdale lies to the right of the photograph, with Batty Moss Viaduct towards the upper left. SD 806774, looking NW.

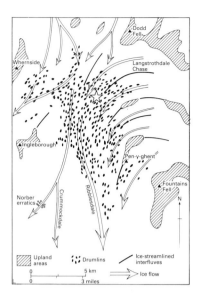

Drumlins in the Ribblesdale area, Yorkshire. Approximate position of photograph 46 is indicated by a star.

to this active erosion, represents an exception to the often-quoted generalisation that uplands are characterised by glacial erosion whereas lowlands are typified by glacial deposition. Here, evidence for both widespread erosion and deposition can be found together.

A very distinctive suite of elongate ridges occupies much of the lower ground in the broad Carsphairn Valley, western Southern Uplands (photograph 47). These ridges reach a maximum height of 20 m about 1 km east of Carsphairn and trend in a direction more or less perpendicular to the valley axis. The photograph shows a portion of these ridges in the Carsphairn Valley.

Studies by R. Cornish showed that these ridges are composed entirely of till, in this case of a type formed at the ice sole.[31] The trend of the ridges across the valley is totally at variance with the down-valley direction suggested by all other indicators of ice flow (e.g. erratics, striae, crag-and-tail landforms): eskers, on the other hand, trend more or less parallel to the last direction of ice flow, which here would have been down-valley. Close inspection by R. Cornish indicated that although the ridges are aligned across the valley, individual crests along the ridges are streamlined or drumlinised in a down-valley (i.e. down-ice) direction.

Cross-valley moraines, like those in the Carsphairn Valley, are more common than their absence from many standard textbooks would suggest. The form of the Carsphairn Valley ridges, their occurrence in a broad bedrock valley, their location close to an ice divide and their close association with drumlins in the

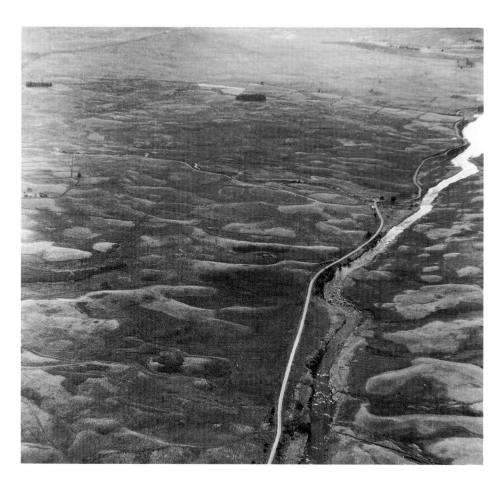

47 Rogen moraine, Carsphairn Valley, south-west Scotland. The Water of Deugh can be seen flowing into Kendoon Loch (upper right). NX 590920, looking SE.

local area are all typical characteristics of Rogen moraine. The drumlinised crests and association with drumlins indicate that these were formed by active rather than by stagnating ice. Formation as a result of over-riding and streamlining of pre-existing ridges by ice has been suggested, but it seems more likely that the transverse elements (the ridges themselves) and longitudinal elements (drumlinised crests) have a common subglacial origin.

FLUVIOGLACIAL LANDFORMS

When large volumes of glacier ice melt, a variety of features form other than those one might recognise from an acquaintance with 'normal' rivers and their deposits. Examples of one of these unusual features are to be found immediately north-east of the outskirts of the built-up area of Wrexham. Photograph 48 shows a series of hollows, most of which are occupied by lakes. They range from a few

48 Kettle holes, NE of Wrexham, Clwyd. SJ 350530, looking NW.

Fluvioglacial landforms near Wrexham. (A) Kames, kettle-holes and sandur development; (B) generalised section through sandur.

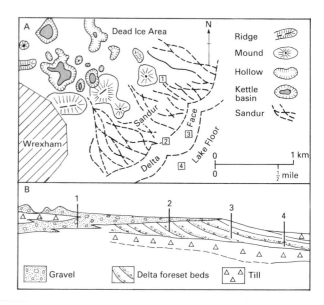

tens of metres to about 500 m across. These hollows are kettle holes and mark the position of former ice blocks which became buried beneath sand and gravel deposits, carried from the margin of the ice mass by meltwater streams. When the buried blocks melted, the overlying sediments collapsed, leaving a hollow. The wastage of a stagnant ice mass present some time between about 18,000 years ago and 14,500 years ago, is thought to have been responsible for the formation of these features. As mentioned earlier in connection with the Ellesmere–Whitchurch moraine, ice moving southwards across the Cheshire Plain from the Irish Sea Basin wrestled for supremacy with ice moving east and north-east from the Welsh Uplands. The junction between the two ice masses appears to have shunted back and forth in an east–west direction approximately along a line joining Wrexham and Mold, as shown by an overlap of tills in the area, one containing erratics from Wales, the other erratics from the Irish Sea Basin. The origin of the kettle holes is linked with that of the so-called 'delta-terrace', a generally gently-sloping surface east and south-east of Wrexham. Various origins have been proposed for this feature, with recently G. S. P. Thomas[32] suggesting that the margins of Irish Sea ice fluctuated leaving a detached stagnant ice mass north-east of Wrexham which gave rise to the kettle holes in the photograph. Meltwater from both Welsh and Irish Sea ice masses became concentrated along the corridor which separated them, leading to a complex interplay of fan development and sandur and deltaic sedimentation in the neighbourhood of the decaying ice mass.

A fine example of the chaotic hummocky terrain that can develop from mainly fluvioglacial deposition associated with the disintegration of Devensian ice is shown in photograph 49. This kame complex occupies an enclave of low land jutting out to the north-east from the floodplain of Cairns Water, Dumfriesshire. In order to interpret this type of landscape, the hummocks and ridges need to be visualised as former sedimentary basins within or on top of a stagnating mass of ice occupying the lower ground, while the intervening low areas represent the positions of former ice cores. As the ice wasted away, the over-steepened sides of the infilled sedimentary basins would have slumped to more stable angles,

49 *opposite* Kame complex, north-east of Cairn Water, Dumfriesshire. NX 870860, looking SW.

dictated by the size of the sediment within them and by the effects of postglacial weathering and erosion. In the photograph, a linear arrangement of elongated hummocks can be distinguished in amongst randomly-oriented hummocks. These elongated hummocks could well represent controlled infilling of individual elongated troughs in a small area of the disintegrating ice mass.

Eskers form near the margin of an ice mass and sometimes present striking features, particularly where they occur in isolation. An excellent example is the Bedshiel esker (photograph 50), located on relatively low-lying ground about 4 km north-north-west of Greenlaw in Berwickshire, south east Scotland. It has the form of a sinuous, symmetrical ridge 3–15 m in height comprising bedded sands and rounded gravels. At its western end it is aligned virtually west–east for about 2 km before turning abruptly northwards and then north-eastwards around the western flanks of Hanged Man's Hill. The change in the direction of the feature, which broadly follows the valley axis, is typical of eskers in that they are aligned

106

roughly parallel to ice flow direction, which locally would have followed the course of the valleys. The esker has several breaks along its length. Since the tops of the esker match in height on either side of some of the breaks, these breaks were probably cut through by meltwater streams after, rather than during its formation at the base of the ice mass.[33]

A whole series of eskers occur in the vicinity of Carstairs, Lanarkshire. Here they form a branching, discontinuous series of linear ridges as much as 24 m high and up to 5 km in length, with numerous less well defined mounds, particularly to the south (photograph 51). The ridge complex represents a classic esker system enclosing numerous kettle holes, while the mounds are considered to be kames. The ridges are aligned in a west-south-west to east-north-east direction in the vicinity of Carstairs itself. All available exposures show that the deposits, comprising sand, gravel, cobbles and boulders, are entirely waterlain, even though some of the boulders in the ridges are in excess of one metre across. These boulders, though, are partly rounded and demonstrate the tremendous power of glacial meltwater streams. The esker system was formed by powerful streams flowing east-north-east in tunnels at the base of an ice sheet.[34] This direction of transport, is indicated, first, by the types of erratic in the system which show an origin to the south-west and, second, by the change in the character of the sediment along the length of the esker system: grain size declines in an easterly and therefore 'downstream' direction, as is the case with 'normal' streams, and there is generally an increased proportion of sand towards the east.[35] The alignment of the system matches quite closely the direction of ice movement of the last ice sheet in the area. Rather unfortunately from the point of view of preserving such a fine example of an esker system, it has proved to be an attractive source of aggregate for the construction industry. Indeed, in Scotland by the early 1980s, a high proportion of all aggregate sources were fluvioglacial in origin. Esker systems in particular are attractive because of the relatively high percentages of rounded gravel that they contain, a testimony to the power of the streams.

While hummocks and ridges characterise fluvioglacial deposits that accumulated beneath, within and on the ice, flat or gently-sloping spreads and terraces typify those laid down against and beyond the ice margin. Terraces are often striking from the air because they have suffered substantial erosion by post-glacial streams cutting down through the outwash deposits leaving steep-sided benches. On the other hand, extensive outwash plains or sandar might easily go unnoticed from the air. This can change where, for example, differences in soil conditions give rise to subtle tonal variations to crops during certain times of the year. This is well illustrated in photograph 52 where the pattern of the last meltwater streams

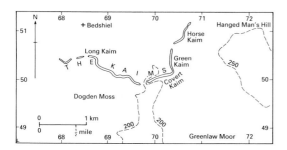

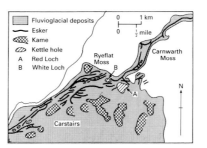

left Bedshiel esker, Berwickshire. Contours in metres.

right The Carstairs esker–kame system, Lanarkshire.

51 Carstairs esker–kame complex, Lanarkshire. Red Loch can be seen at the centre of the photograph and White Loch to the left of it. Carstairs itself can just be seen beyond White Loch. NS 958474, looking WSW.

to flow across a former outwash plain near Brechin, eastern Scotland can be seen. The photograph shows just a small part of a sandur spread out across low ground by the glacial forebears of the present rivers North Esk and West Water. The nature of the outwash changes in coarseness from predominantly cobbles and boulders near the uplands to sand and small gravel in the vicinity of the view shown.

Upstream towards the mouths of Glen Esk and the West Water glen, the outwash is replaced by kame and kettle topography which marks the stagnating margin of the ice from which issued the meltwater streams that built up the sandur. The photograph depicts well the criss-crossing of the outwash by a maze of shallow, broad streams, many of which today have little surface morphological expression. The tendency for meltwater streams to divide and flow in a number of shallow channels (i.e. to 'braid') is typical of such streams and is caused primarily by rapid changes in the amount of water issuing from the glacial margin and by the ready availability of easily eroded material for transport in the largely unvegetated proglacial environment. With the disappearance of glacier ice and the onset of relatively consistent flow regimes in the area, combined with a stabilising plant and soil cover, the West Water and North Esk have become confined to single meandering channels incised into the surrounding sands and gravels deposited by their glacial forebears.

Farther south, in the Forth Valley, east-central Scotland, lies another large expanse of fluvioglacial deposits. Photograph 53 shows part of a large outwash delta in the Larbert-Falkirk area. The delta was much dissected by later fluvial erosion and in recent years the extent of surviving remnants has been reduced further by the sand and gravel extraction industry. The delta merges eastwards with raised beach deposits formed at a time when sea level recovery following maximum glaciation was comparatively well advanced with the sea penetrating a long way up the Forth Valley which, like much of northern Britain, remained considerably depressed due to downwarping caused by the tremendous weight of the ice sheets. Upvalley the delta deposits give way to kames, kame terraces and kettle holes, indicating the former presence of a stagnating ice mass. For a number of years up until the early 1970s, this large expanse of fluvioglacial deposits was linked with other similarly large sequences of meltwater deposits

52 West Water, Angus. Crop marks pick out the braided channel patterns across a former outwash plain. NO 617657, looking N.

53 Camelon, east-central Scotland. Sand and gravel extraction taking place in a dissected remnant of an outwash delta. The delta was formed by meltwater streams flowing into the sea during the retreat of last glaciation ice sheets, some time between about 18,000 and 13,000 years ago. NS 863811, looking SE.

in Scotland, including the Carstairs esker system and the outwash plain near Brechin featured in the previous photographs, to support the now-rejected Perth Readvance.[36] Once some key evidence for the readvance had been refuted and no other firm evidence in support of it was forthcoming, fluvioglacial sequences, like those in the Forth Valley and in other areas in Scotland were reinterpreted instead as the products of a major glacial stage.

With such great quantities of meltwater being produced in the neighbourhood of masses of stagnating ice and retreating glacier margins, one might suppose that temporary lakes would have formed. The question might then be raised as to what types of evidence would indicate the existence of a former ice-dammed lake. Certainly the former existence of ice-dammed lakes has been of interest to glacial geomorphologists for over 100 years and a large number of such lakes has been proposed for many areas of Britain. Much of this interest may be traced back to P. F. Kendall who wrote about 'a system of glacier-lakes in the Cleveland

Hills' in 1902.[37] His main evidence for these lakes was the so-called 'spillways' or 'overflow channels' by which the water would have drained from the lakes. Later detailed mapping indicated that in some of these spillways the bedrock floors of the channels actually rose and fell along their length by several metres indicating that they could not have been produced by 'normal' streams. Having established the spurious nature of the chief evidence, J. B. Sissons called for the critical appraisal, in articles written in 1960 and 1961,[38] of many of the former ice-dammed lakes, and associated 'overflow channels' scattered around Britain, which were accepted generally at that time. He proposed a number of features which should be found in the area around a former ice-dammed lake. These included evidence of shorelines, deltas formed as the velocity of streams entering a lake was checked and lake floor deposits comprising laminated sediments. As a result of the publication of these articles, it became appreciated rapidly that many of the proposed lakes could no longer be accepted and other explanations for the 'overflow channels' were sought, usually involving the idea that they had formed subglacially or in a position marginal to the ice. Nevertheless, the evidence for some proposed ice-dammed lakes remained compelling (e.g. Glen Roy and its 'Parallel Roads', north west Scotland) and an ice-dammed lake origin is still accepted. New evidence has also been amassed for previously unknown former ice-dammed lakes. The mosaic of three vertical air photographs (photograph 54) shows some of the depositional evidence supporting the former

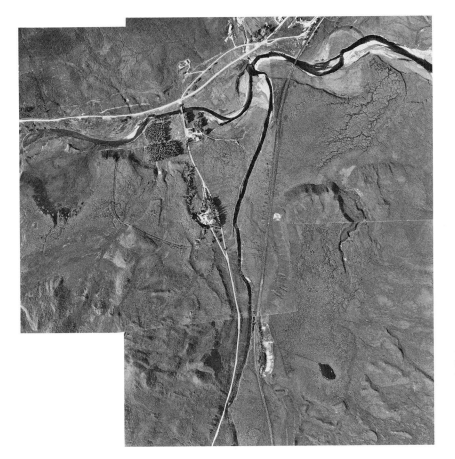

54 Achnasheen, central Ross-shire. A mosaic of three vertical aerial photographs shows evidence for a former ice-dammed lake in Strath Bran. See the diagram on p. 112 for an interpretation of this mosaic. NH 170590.

existence of an ice-dammed lake near Achnasheen, central Ross-shire. The most striking evidence consists of two high delta terrace fragments that have steep frontal slopes up to 20 m high on the valley floor opposite the main road passing through the hamlet of Achnasheen (see below and diagram on p. 113).

These delta terraces were formed during the retreat of the last ice sheet by meltwater from the snouts of two ice masses where the streams entered an ice-dammed lake in Strath Bran. One of the ice masses moved from the direction of Loch a' Chroisg west of Achnasheen and the other from Loch Gowan to the south which reached the upper ends of the terrace fragments as is indicated by ice marginal features such as kame and kettle complexes and steep ice-contact slopes to the upvalley ends of the terraces. The eastern end of Strath Bran was blocked by an ice mass that reached a point on the valley floor some 3 km east of Achnasheen and this is marked by a large end moraine. The fronts of the two delta terraces match closely in height at about 175 m, suggesting that they were formed at the same time, since this lake level was only one of several levels reached during what must have been a relatively short period post-dating the Wester Ross Readvance.[39] This 175 m level was maintained by the height of a bedrock overflow that discharged subglacially at the eastern end of the lake. The lake level then fell to about 140 m, resulting in series of lower terraces being produced by the meltwater streams issuing from the two ice tongues at the western end of Strath Bran. The lake was by now dammed by the end moraine which in post-glacial times became eroded by the River Bran.

(A) Glacial and fluvioglacial landforms near Achnasheen; (B) Inferred glacial limits at the time of maximum development of ice-dammed lake; (C) simplified section through the delta terraces and floodplain.

55 Achnasheen high terrace from the ground.

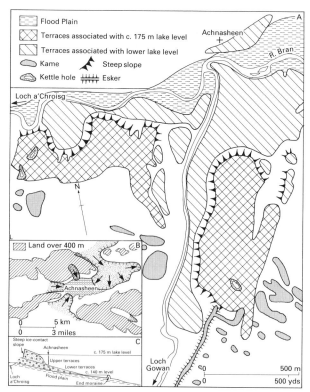

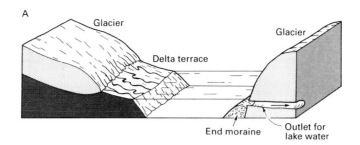

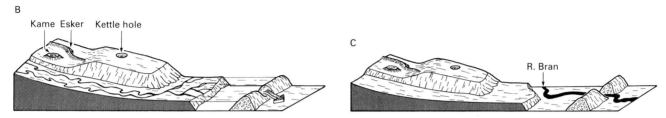

Of the 20 reproductions of aerial photographs in this chapter, 18 are oblique views of depositional landforms and in one case (photograph 35) of a particular source of distinctive glacial erratics found in glacial deposits around the Irish Sea Basin. These 'bird's eye' views give an excellent opportunity for expert and layman alike, particularly for lowland areas where no high vantage point is otherwise available, to appreciate more fully a pattern to what otherwise might seem a chaotic arrangement of hummocks and ridges. The impact of such photographs is immediate because the viewpoint is just like that from a hill or mountain top. Although only two reproductions of vertical aerial photographs are given in this chapter (photographs 42 and 54), nevertheless for the geomorphologist investigating glacial and fluvioglacial landforms, vertical aerial photographs would be preferred even though their interpretation takes longer and their visual impact is usually less immediate. There are three main reasons for this. First, overlapping vertical aerial photographs can provide a three-dimensional image when viewed through a stereoscope, which greatly increases their value in interpreting even very small landforms. Second, the information on a vertical aerial photograph is relatively easy to transfer to a large-scale map without the complicated calculations needed if oblique aerial photographs were used for this purpose. Third, dimensions of landforms are obtainable directly from a vertical aerial photograph.

Simplified model of the development of features associated with the former ice-dammed lake, Achnasheen. (A) Ice-dammed lake at 175 m level; (B) at 140 m; and (C) at the present day.

GUIDE TO FURTHER READING

K. Addison, *Classic Glacial Landforms of Snowdonia. Landform Guide No. 3*, Sheffield, 1983.

C. Embleton and C. A. M. King, *Glacial Geomorphology*, London, 1975.

B. S. John, *The World of Ice*, London, 1979.

R. J. Price, *Glacial and Fluvioglacial Landforms*, London, 1973.

R. J. Price, *Scotland's Environment during the last 30,000 years*, Edinburgh, 1983.

B. W. Sparks and R. G. West, *The Ice Age in Britain*, London, 1972.

D. E. Sugden and B. S. John, *Glaciers and Landscape: A Geomorphological Approach*, London, 1976.

4 Periglacial landforms

Beyond the edge of the glaciers during the Quaternary glacial periods in Britain, climatic conditions akin to the modern High Arctic prevailed. These periods of periglacial climate have left their mark on the landscape, and in many cases the resulting landforms survive to the present day. By studying the climatic conditions and geomorphological processes in the modern periglacial zone, geomorphologists are able to explain the nature and origin of the periglacial landforms of Britain, and to use them to reconstruct past climates. For detailed accounts of periglacial environments and landforms the reader is referred to the text books by H. M. French[1] and A. L. Washburn.[2] Today periglacial climates occur both at high latitude, in the Arctic and the ice-free areas of the Antarctic, and in mountainous areas at lower latitudes where high altitude results in low average temperatures.

French defined an absolute limit of $+3°C$ mean annual air temperature for the modern periglacial climatic zone. He suggested that from a geomorphological point of view, areas with mean annual temperatures below $-2°C$ are fully periglacial, in that processes associated with frost action predominate, while areas with mean annual temperatures between $-2°C$ and $+3°C$ are marginally periglacial since other geomorphological processes not associated with frost are likely to dominate.

Three major subdivisions of the modern periglacial realm may be identified, although each includes a range of climatic conditions. Beyond the Arctic and Antarctic tree line is the tundra zone, which gives way to polar desert in the High Arctic and in ice-free areas of the Antarctic mainland. Here the environment is characterised by continuous permafrost (permanently frozen ground), short summers and long, severe winters. The Subarctic boreal forest forms the second periglacial zone, with climates ranging from continental extremes of winter cold and summer warmth, to much more equable maritime conditions. The boreal forests lie mainly in the discontinuous permafrost zone, with mean annual temperatures ranging from around $-1°C$ to $-6°C$. The third major periglacial climatic type corresponds to high altitude areas at lower latitudes, sometimes referred to as 'alpine' periglacial regions. Here, in the alpine tundra above the tree-line winter temperatures are generally less severe than in the Arctic tundra, but short-term, often diurnal, freezing and thawing cycles are frequent, especially in spring and autumn. Precipitation is also much higher than in the other two regions, with considerable winter snow accumulation.

Periglacial landforming processes depend to a greater or lesser extent on low temperatures which promote freezing of water in winter, and warmer temperatures which lead to some thawing in summer. Freezing and thawing causes accelerated physical weathering of bedrock and the development of a range of distinctive landforms associated with frozen ground. Some processes, notably those associated with permafrost, are unique to periglacial areas, while others, such as physical weathering of bedrock and mass movement processes on slopes, are sufficiently accelerated under periglacial conditions to produce distinctive landforms.

The operation of running water as an agent of erosion and deposition is important in all humid environments, including most periglacial ones. However, the hydrology of periglacial rivers differs markedly from that of rivers in other climates, since the majority of the winter precipitation is stored up as snow. This accumulated winter snow is released rapidly during the spring and early summer thaw, producing a characteristic 'nival flood'. Later in summer, following the flood, discharges fall considerably, while in winter water is frozen and rivers cease to operate.

Another process not restricted to periglacial areas but which is sometimes of importance, is wind action. As in hot deserts, a lack of vegetation cover, low precipitation and the availability of fine-grained sediments at the ground surface, facilitate deflation and aeolian transport of silt and fine sand in some periglacial areas. Wind action has been shown to be important in such diverse periglacial environments as southern Victoria Land (an ice-free area of polar desert in Antarctica), West Greenland, and the Cairngorm Mountains of the Scottish Highlands. During Quaternary glacial periods wind action was particularly effective in the dry continental periglacial climates beyond the ice sheets. In Europe, considerable thicknesses of 'loess' (wind-blown silt) were deposited by dry easterly winds in a great swathe running from Russia to western France. Britain lay on the north-western margin of this loess belt and loessic soils are present in southern and eastern England and the extreme south of Wales.[3]

Probably the most important periglacial phenomenon with regard to land-forming processes is frozen ground. Frozen ground may be subdivided into two classes: seasonally frozen ground which thaws annually; and permafrost, which remains permanently frozen.[4] Although such processes as periglacial mass movement of soil on slopes and frost sorting of near-surface sediments may be promoted by seasonally frozen ground, it is permafrost which is directly or indirectly associated with many of the geomorphological processes acting in modern periglacial areas. The term permafrost is defined purely on the basis of temperature, irrespective of the presence of frozen water. Thus, ground where the temperature has remained below 0°C for more than two years is deemed to constitute permafrost.[5] It is, however, the presence and distribution of ice within permafrost which is of major geomorphological importance.

Wherever mean annual ground temperature falls below 0°C, the depth of winter freezing will exceed the depth of summer thawing and a layer of permafrost will grow downwards into the ground. The permafrost layer will thicken each winter until equilibrium is reached between the flow of heat upwards to the atmosphere from the base of the permafrost, and geothermal heat flow to the base of the permafrost from the centre of the earth, as shown on the graph. In northern Alaska permafrost extends to depths of around 650 m, and in northern Siberia it is around 1,600 m thick. As climate becomes less severe towards the south so

Hypothetical temperature
profiles in permafrost,
representing conditions in
central Alaska.

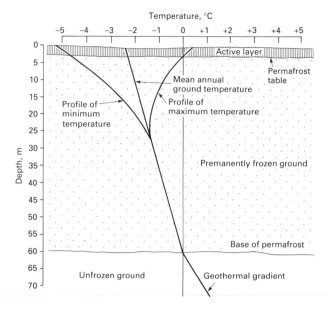

the permafrost thins progressively. The surface layer which thaws during summer is called the 'active layer' and its thickness varies from a few tens of centimetres in the high latitude polar deserts to 2 m to 3 m near the limits of discontinuous permafrost.

Field studies in North America and Siberia have shown that where mean annual temperatures are below $-6°C$ to $-8°C$, permafrost will be continuous. However, where the climate is less severe (mean annual temperatures between $-1°C$ and $-6°C$), the insulation provided by vegetation, winter snow cover, small ponds, peat bogs etc. leads to areas where permafrost fails to develop. Since these insulating factors are spatially highly variable, a complex pattern of discontinuous permafrost results. In alpine periglacial regions permafrost is present at higher altitudes. Discontinuous permafrost has been observed in the Scandinavian mountains where mean annual air temperatures are below $-1.5°C$ and continuous permafrost where mean annual air temperatures are below $-6°C$.[6]

The identification of these climatic indices for permafrost is of considerable value in the reconstruction of Quaternary periglacial climates in countries such as Britain where temperate climates now prevail. As we shall see, a number of landforms develop only in areas underlain by permafrost. Where these can be recognised in the landscape, or as structures disturbing sedimentary sequences, their presence provides vital evidence for the former presence of permafrost, from which inferences may be drawn about mean annual air temperatures.

PERIGLACIAL BRITAIN

With regard to its Quaternary history, Britain may be subdivided into three main areas, as shown on the following map: Region 1 to the south of the Quaternary glacial maximum; Region 2, between the limits of the last glaciation (the Devensian) and the Quaternary glacial maximum; and Region 3, to the north of the Devensian ice limit. In Region 1 periglacial climates prevailed during all the Quaternary cold periods, which for the last 500,000 years at least, may account

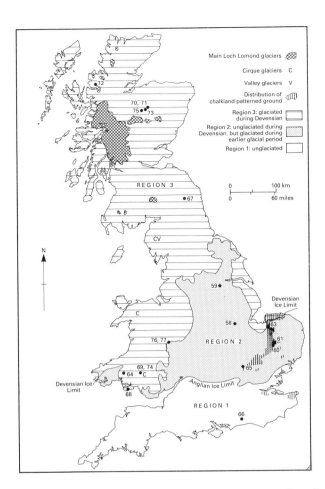

The Quaternary in Britain,
showing the three zones
defined in the text, and
locations of aerial photographs.

for 80 per cent of the time. Region 2 was ice-free during the Devensian Glacial
(approximately 90,000 years to 10,000 years ago) and therefore suffered prolonged
periglaciation, but was glaciated during at least one of the earlier Quaternary
cold periods. In Region 3 although severe periglacial climates prevailed during
the early and middle Devensian, many of the resulting landforms were subsequently destroyed or obscured by glaciation in the late Devensian. Even here,
however, periglacial conditions returned immediately following clearance of the
glaciers, and were certainly re-established by the sharp deterioration in climate
which occurred around 11,000 years ago. This was the Loch Lomond Stadial,
which persisted for about 1,000 years, and led to the development of a small
ice cap in the Scottish Highlands (see the map above), corrie glaciers in other
mountainous areas of Scotland, England and Wales, and permafrost as far south
as East Anglia.

Climatic amelioration was rapid after 10,000 years ago, and temperate conditions have prevailed since then. However, it should be remembered that the
higher Scottish mountains, and to a lesser extent the mountains of the Lake District
and North Wales, are significantly colder and wetter than the lowlands. Ben
Nevis, for instance, at 1,343 m Britain's highest mountain, has a mean annual
temperature only fractionally above 0°C at its summit. A maritime periglacial
climate in the British mountains has been identified, which is characterised by
high precipitation, prolonged snow cover, and frequent strong winds.[7]

Since permafrost is widespread in the modern periglacial zone, and of major geomorphological significance, its effect on the periglacial landscape of Britain will be considered first. In areas with bedrock at, or very close to, the surface the presence of permafrost is often of little significance, but where unconsolidated sediments form the substrate, perhaps laid down by a glacier or a river, permafrost may be associated with large volumes of clear ice within the frozen sediment. Such 'ground ice' (excluding the ice occupying pore spaces between the sedimentary particles) has been classified into three major groups; ice-wedge ice, segregation ice, and injection ice, according to the mode of ice accumulation.[8] The effect of these three types of ground ice on periglacial landforms is distinctive, and may still be detected in the British landscape, especially when viewed from the air.

FOSSIL ICE-WEDGE POLYGONS

Ice-wedge ground ice consists of vertical or near vertical wedge-shaped veins or dykes of ice between 1 cm and 3 m wide penetrating up to 10 m into the permafrost[5] (photograph 56). In plan the linear wedges link up to form polygonal networks marked at the surface by troughs in the active layer (photograph 57). Individual polygons may range in diameter from 3 m to 30 m or more. The plan form may be roughly hexagonal with angles between the sides tending to approximate 120°, or orthogonal with a tendency towards 90° junctions. Hexagonal polygonal patterns are considered to indicate a more or less synchronous development of cracks, while the orthogonal patterns result from the initial formation of random primary cracks which are followed by secondary cracks which progressively subdivide the area and develop orthogonal intersections.[9]

The formation of ice-wedge polygons results from severe cooling of the permafrost in winter. If ground temperatures fall to around −15°C or −20°C, thermal

56 Top of an ice wedge exposed by placer mining in river gravels, Hunker Creek, Klondike District, Yukon Territory, Canada. The wedge is approximately 2.5 m wide at the top.

contraction of the ground generates sufficient stress to cause cracking. These stress relief cracks form a polygonal pattern, rather like mud cracks around a drying pond, but on a much larger scale. Average crack widths of around 1 cm have been recorded in the Mackenzie Delta, Canadian Arctic.[10] Snow may enter these open cracks, and hoarfrost may form in them, so that as the permafrost warms in the spring and summer the resulting thermal expansion is unable to completely close the cracks, and thin veins of ice are left penetrating downwards, marking the former crack positions. Cooling during subsequent winters will again cause the cracks to reopen and snow and hoarfrost will tend to fill them, so that a yearly increment of ice is added to each wedge, which thereby grows slowly thicker. In Alaska and in the Mackenzie Delta, the annual ice-wedge growth is only a few mm at most, and often less than 1 mm, so that thick ice wedges must take many hundreds or even thousands of years to form.[11]

Amelioration of the climate, leading to thawing of the permafrost, results in slumping of surface sediments into the voids left by the melting ice wedges, so preserving their forms as 'ice-wedge casts' within the host sediments (photograph 58). Such ice-wedge casts are frequently encountered in Quaternary sediments in Britain and when their stratigraphic relationships are known, may be used to identify periods when permafrost prevailed during the Quaternary.[12]

Since the sediment which fills the ice-wedge casts often differs in texture from the surrounding host material, the polygonal pattern formed at the surface by the sediment-filled former ice wedges may sometimes be detected as crop marks. These result from unequal growth of crops or differences in moisture status along the surface outcrop of the ice wedge casts, and under favourable conditions may be clearly seen from the air. The example shown in photograph 59 has a random orthogonal pattern with a series of relatively wide and continuous fossil ice wedges and many narrower less continuous links which form an irregular crazed pattern on the ground surface. Careful examination reveals a larger poly-

58 Two ice-wedge casts, Baston, Lincolnshire. The ice wedges developed in fluvial gravels probably in the Late Devensian, and penetrated the underlying Oxford Clay. Two generations of ice-wedge development are shown, the smaller cast on the left is earlier, since it is truncated at the top by the larger cast.

59 Fossil ice-wedge polygons, near Adwick-Le-Street, Yorkshire. The polygons are developed in sand and gravels overlying Magnesian Limestone. Note the random orthologonal pattern, with larger, more continuous fossil ice wedges forming large-scale polygons which are subdivided into smaller polygons by narrower, less continuous secondary wedges. SE 534048, looking SW.

gonal network with polygon diameters 20 m to 50 m, which is subdivided by minor fissures into much smaller polygons 5 m to 10 m in diameter. This pattern strongly suggests sequential development, and it has been suggested by Russian investigators that the formation of such smaller polygons which subdivide larger ones results from increasingly severe winter temperatures.

This example from Yorkshire falls in periglacial Region 2 of the British Isles (see the map on p. 117), that is, the area beyond the Devensian ice limits but inside the maximum of earlier Quaternary glaciation. Similar fossil ice-wedge polygons, but with a hexagonal pattern, occur in a terrace of the Worcestershire Avon.[13] This area was also beyond the Devensian ice and it may be concluded that the most likely period of polygon formation was during the coldest part of the Devensian, contemporaneous with ice advance. However, near Wolverhampton, fossil ice-wedge polygons occur in Devensian till. Here hexagonal patterns show average diameters ranging from 2.8 m to 6.2 m and excavation has confirmed the presence of clearly defined ice-wedge casts below the polygon boundaries.[14] Since permafrost and ice-wedge development must have post-dated ice retreat, the most likely time of formation here was immediately following deglaciation, that is, between approximately 13,500 BP and 12,500 BP. Farther north, in Scotland, deglaciation took place somewhat later and climatic amelioration following ice retreat was rapid. This suggests that polygonal crop markings observed in North-East Scotland for instance, result from ice-wedge polygons forming in permafrost during the Loch Lomond Stadial, approximately 11,000 BP to 10,000 BP.[15]

LARGE-SCALE PATTERNED GROUND

The second type of ground ice associated with permafrost is segregation ice. This consists of small lenses of ice a few mm to a few cm in thickness formed during freezing of fine-grained sediments. Due to molecular forces between the sediment particles and the pore water, the freezing point of water in very fine pores is lowered to below 0°C. During ground freezing therefore, water may migrate into the sediment immediately behind the 0°C isotherm, where the temperature is slightly below zero, and freeze on to growing ice lenses. If water may be drawn readily from the unfrozen ground, perhaps fed by groundwater, a considerable volume of ice is incorporated in lenses, and the ground above is heaved upwards to accommodate the increased volume of the resulting frozen soil. Ice segregation may take place during permafrost development, producing ice-rich permafrost, and also annually in the active layer as it refreezes in winter.

It is generally considered that variations in the rates and directions of ice segregation in the active layer were responsible for the formation of a second type of fossil patterned ground found in Britain, which occurs in the chalklands of East Anglia and the Isle of Thanet, Kent.[16] Again the features are picked out as crop marks due mainly to variations in vegetation (photographs 60 and 61). They are variable in plan, ranging from cellular nets to U-shaped or doughnut-shaped features (photograph 60). Where the gradient exceeds 1°–2° they become elongated downslope before merging into a striped pattern (photographs 60 and 61). The diameter of the nets is generally around 10 m, and the spacing of stripes 7 m. The patterns are developed where weathered chalk is mantled by sandy,

possibly wind-blown sediments. Sometimes, as at Thetford Heath, Norfolk, chalky till underlies this sand, rather than weathered chalk.

Excavation reveals an up-doming of the chalky substrate beneath the centres of the nets, with only a thin cover of sand (see diagram A below) which is mixed with fine chalky loam and is in consequence calcareous. In contrast, beneath the margins the sand is up to 1.5 m thick, noncalcareous, and of coarser texture. Chalkland grasses tend to cover the calcareous centres, while heather (*Calluna vulgaris*) covers the noncalcareous sandy margins. Sections excavated across striped patterns show them to be virtually identical to the polygonal forms with calcareous, grass-covered stripes alternating with sandy *Calluna*-covered stripes. The presence of vertically aligned chalk rubble and flints has been recorded beneath the chalky centres of the polygons and stripes, with stones tending to lie horizontally beneath the sandy troughs. In many sections tongues of sediment penetrate upwards into the chalky centres, forming involutions. These involutions are much smaller than the patterns themselves, usually less than 1 m across.

A wide variety of patterned ground phenomena, thought to result largely from ice segregation processes in the active layer, occur throughout the modern periglacial zone. These patterns have been classified into sorted forms, where the pattern is picked out by variations in sediment grain size, and non-sorted forms where micro relief and vegetation are the main variables.[17] We shall deal with sorted features later in this chapter; it is with large scale non-sorted forms, which may provide analogues for the chalkland patterned ground, that we are concerned here.

Non-sorted circles are common in the subarctic boreal forests and the arctic tundra. On horizontal surfaces in the Seward Peninusla, Alaska, they have diameters of 6 m to 9 m, with slightly domed centres and trough-like margins (diagram B below).[18] Where the ground is slightly sloping the circular patterns

60 *opposite: top* Large scale non-sorted chalkland patterned ground, near Elveden, Suffolk. The polygonal pattern on the gently inclined hill summit in front of the copse becomes elongated and passes into stripes as the gradient increases in the foreground. Behind the copse a similar transition from polygons to stripes is apparent. TL 828780, looking SW.

61 *opposite: bottom* Large scale non-sorted chalkland patterned ground, Thetford Heath, Suffolk. Again a transition from polygonal networks on hill crests to stripes on valley sides can be seen. TL 845800, looking NW.

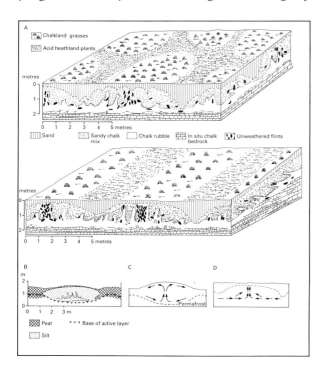

(A) Representative diagrammatic sections through Chalkland circles and stripes.

(B) Section through non-sorted circle, Seward Peninsula, Alaska.

(C) Postulated pattern of sediment movement in non-sorted circles due to freezing and thawing of the active layer.

(D) Pattern of sediment movement in chalkland patterned ground suggested by concentrations and alignment of flints.

become elongated downslope, and give way to stripes formed of ridges and fur-rows. In Canada such non-sorted circles are referred to as earth hummocks or simply hummocks.[19] A slow circulatory movement of sediment is envisaged to take place within these features, set up as a result of ice segregation along a sinuous freezing plane (diagram C above).[20] During the thaw, water percolates down the sides of the bowl-shaped permafrost table below the hummock and refreezes, forming ice lenses, and causing upward heave of the hummock centre. Thawing of this ice-rich sediment beneath the hummock centre towards the end of the summer produces a semi-liquid mud which, due to its lower density, tends to rise upwards from the permafrost table into the centre of the hummock.

The similarity in surface and subsurface form of the fossil chalkland patterns to the nonsorted patterns described from the subarctic and arctic is clear, and it seems likely that the chalkland patterned ground of eastern England developed during the Devensian when permafrost was present. Over 500 localities with this type of patterned ground are known from Norfolk and Thanet, but further west, even on chalk areas, such patterns are absent. Possibly the more continental climate of eastern England compared with that of the west may have been import-ant, but the presence of fossil ice-wedge polygons further west, especially in the Midlands, indicates that permafrost was present at the time that the chalkland patterns were most likely to have formed. To conclude by analogy with probable present day equivalents, the chalkland patterns mark a period of continuous permafrost in eastern England, with active layer depths of around 1.5 m.

FOSSIL GROUND ICE DEPRESSIONS

The third major class of ground ice in permafrost is injection ice, which occurs as lenses or sheets of ice several metres in thickness and in some cases several square kilometres in area. During freezing of saturated sands and gravels water may be expelled in front of the advancing freezing plane. This expelled water collects at or near sedimentary boundaries, perhaps above a clay layer, where it freezes to form injection ice. A particular form of injection ice is associated with conical ice-cored hills called pingos. In this case water freezes within a few metres of the ground surface to form a blister-like ice mass, pushing up the ground surface into a pingo (photograph 62). Two distinct types of pingo have been identi-fied; so-called closed system pingos, formed by refreezing of unfrozen ground beneath drained ponds and lakes in the continuous permafrost zone,[21] and open system pingos, which occur in the discontinuous permafrost zone as a result of freezing of water percolating laterally beneath thin permafrost[22] (diagram A below). Open system pingos occur mainly on the lower parts of slopes, where groundwater freezes as it seeps upwards, under artesian pressure, into the base of the permafrost. The ground surface is pushed upwards into an ice-cored hill. Eventually, disruption of the vegetation and soil as the pingo grows leads to exposure of the ice core at the surface (photograph 62) and melting. The feature then gradually collapses as its ice core melts, to form a circular pond surrounded by a rampart-like ridge of sediment pushed up during pingo growth. In Alaska open system pingos have diameters between 15 m and 440 m, with the majority falling in the range 150–185 m.[23]

62 Small closed-system pingo, Mackenzie Delta, North-West Territories, Canada. Tree trunks deposited by the nival flood give an impression of scale. This pingo has similar dimensions to the open system pingos of the discontinuous permafrost zone. Note rupture of the surface exposing the pingo ice core. This will lead to thawing and pingo collapse.

Similar ice-cored mounds may also develop due simply to localised ice segregation. Where peat-covered, these features are called palsas, and where they are formed in mineral soils they have been termed 'mineral palsas'.[24] Melting of mineral palsas also leaves small circular ponds surrounded by low ramparts of mineral soil. In fact there is a continuum between open system pingos and mineral palsas, but in general the former are likely to develop where groundwater seepage takes place, usually on the sides or in the bottoms of valleys, while the latter form on plateaux where no such seepage can occur.

In several places in England and Wales the remnants of collapsed open system pingos and mineral palsas can be seen as circular or elongated depressions, sometimes surrounded by low ridges (photographs 63 and 64). Photograph 63 illustrates perhaps the best known of these ground ice features, at Walton Common, Norfolk.[25] A section provided by a gas pipeline trench showed that the depressions concealed smooth-floored basins about 7 m in depth filled with peat and organic silts (see diagram A, p. 126). The section also revealed two distinct periods of activity, with an older, larger basin containing a layer of organic silts covered

(A) Typical location and hydrological conditions of open system pingos in central Alaska.

(B) Hydrological conditions at Walton Common.

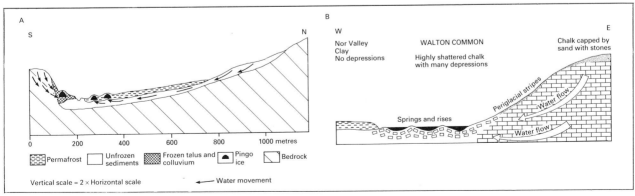

63 Fossil ground ice depressions, possibly open system pingos, Walton Common, Norfolk. The near circular ramparts in the centre and centre-right of the photograph mark the rims of former ice-cored mounds. Irregular ridges result from superposition of successive generations of mounds. TF 735165, looking S.

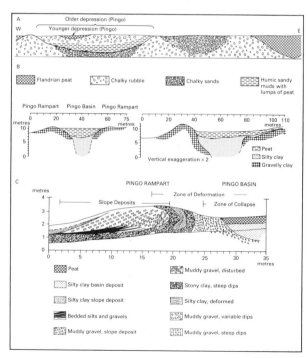

(A) Section through southern part of Walton Common showing two generations of ground ice depressions.

(B) Examples of sections through fossil pingos in the Cledlyn Valley, central Cardiganshire.

(C) Section of pingo rampart, Cletwr Valley, central Cardiganshire.

by soliflucted chalk and a smaller peat-filled basin, developed in the chalky solifluction fill. Pollen analysis showed that the organic silt band in the bottom of the older basin began to accumulate at the beginning of the Windermere Interstadial (Pollen Zone I/II boundary), while the peat filling the later basin began to accumulate at the beginning of the Post-Glacial (Pollen Zone III/IV boundary). It was concluded that pingos and palsas began to develop in East Anglia during the Devensian, through to Pollen Zone I but that a second generation was formed under discontinuous permafrost during the Loch Lomond Stadial (Pollen Zone III; 11,000–10,000 BP).[26]

The hydrological situation at Walton Common was extremely favourable to open system pingo formation, with a ground water supply likely to lead to injection ice growth (see diagram B on p. 125). At some other sites, on gentle interfluves for instance, this may not have been the case and the ground ice mounds responsible for depressions may have been of the mineral palsa type.

A second group of ground ice depressions is found in west and south-west Wales where they occur in valley bottom and lower valley side locations. The parent material is generally solifluction sediments. The example shown in photograph 64 is from a tributary valley of the Afon Gwili in Carmarthenshire and shows identical features to those in Cardiganshire, where basins are 30–40 m in diameter and ramparts about 20 m wide and up to 3 m high.[27] Excavation of a rampart in the Cletwr valley, Cardiganshire, showed clear evidence for compres-

64 Fossil ground ice depressions (open system pingos or mineral palsas), SW of Llanpumpsaint, Carmarthenshire. Note the small circular depressions, right-centre of the photograph and beyond the road. The largest feature, adjacent to the road has a broad elongated depression surrounded by a U-shaped ridge. SN 417274, looking SE.

sion of the sediments around a growing ice core (see diagram C on p. 126) and slumping of sediment on the outer flanks. Augering of several basins showed them to be peat-filled hollows up to about 5 m in depth, but in the Cledlyn valley, Cardiganshire, a superficial peat layer up to 3 m thick covered silty clay sediments filling a hollow more than 10 m deep (see diagram B on p. 126).

Repeated formation and collapse of ground ice mounds produced a high density of ramparts and basins and several composite features. The location of these depressions at the foot of long slopes is similar to the modern open system pingos of central Alaska, and the deformation of sediments incorporated in the excavated rampart also suggests that injection ice was responsible for the ground ice mounds. It has recently been suggested, however, that these 'fossil pingos' in South-West Wales may be the remains of 'mineral palsas', since the supply of groundwater through the low permeability slope materials may have been insufficient for injection ice to develop.[28]

Radiocarbon dating of organic material in the depressions shows that they began to fill around 10,000 years ago, that is, towards the end of the Loch Lomond Stadial. Unlike the East Anglian features, those in Wales are largely inside the ice limits of the last glaciation, so that they must have developed under conditions of discontinuous permafrost in the period immediately following deglaciation (Pollen Zone I) and again during the Loch Lomond Stadial (Pollen Zone III), up to about 10,000 years ago.

CHALKLAND DRY VALLEYS

The high porosity of chalk renders it a relatively permeable rock. In Britain, under present-day climatic conditions, rainfall is generally not of high enough intensity to generate surface runoff. Instead precipitation percolates into the rock, seeping down to the groundwater table. Where the water table intersects the ground surface, however, groundwater springs occur which feed the chalk streams below. Despite the lack of surface streams at higher altitudes, the chalk of southern England is dissected by many dry valley systems which clearly are not forming under the present climate.

It was Reid in 1887 who first suggested that these dry valleys or 'coombes' formed under periglacial conditions when the chalk was rendered impermeable by the presence of permafrost.[29] Meltwater, he argued, was able to run off over the surface, eroding valleys which subsequently became dry when impedence of drainage by permafrost ceased. There is, as we shall see, considerable evidence that these valleys developed under a periglacial regime, and that they have changed very little during the post-glacial period. However, several authors have pointed out that most chalk dry valleys are almost certainly polycylic in origin, having probably been initiated during periods of higher water table and enlarged and greatly modified under periglacial conditions.[30]

On scarp faces coombes are often steep sided, extending downslope from paddle-shaped amphitheatres cut into the escarpment (photograph 65). In some cases, such as Devil's Dyke, near Brighton, they form spectacular trenches incised up to 100 m into the chalk. On the south-facing escarpment of the North Downs in Kent the coombes have steep smooth headwalls and in the larger examples

flat floors which widen downslope, opening out on to the Gault Clay vale. An apron of chalk muds together with angular chalk rubble is often present, extending from the mouth of the coombe over the surface of the clay vale.

In the Devil's Kneadingtrough, near Brook, Kent, much of the valley fill and apron consists of fine mud deposited by surface wash.[31] Dividing the colluvium sequence is a buried soil of Windermere Interstadial age (Pollen Zone II) showing that erosion of the chalk and deposition of valley bottom sediments by meltwater took place during the Devensian, but was interrupted by soil formation on a stable surface during the short-lived Windermere Interstadial. Renewed meltwater activity in the cold Loch Lomond Stadial then led to burial of this soil.

Underlying the surface wash deposits at Brook, and present in virtually all the chalk dry valleys of southern England are deposits comprising angular lumps of chalk set in a fine powdery chalk matrix. This material, called 'coombe rock' is the result of frost shattering of the chalk and subsequent sludging of the shattered rock as a saturated mass down into the valley bottoms by the process of gelifluction. Measurements of the orientation of elongate chalk clasts within the coombe rock show a consistent downslope alignment, parallel to the valley side slopes, a clear indication of downslope mass movement prior to sediment accumulation.

65 Dry valley in chalk escarpment, Barton in the Clay, Bedfordshire/Hertfordshire. TL 088296, looking SSE.

129

66 Dry valley in chalk, SE of Amberley, Sussex. The valley shows asymmetry, but largely as a result of valley meanders, with steeper slopes on the concave sections on the outside of meander bends and gentler slopes on convex spurs on the inside of valley meander bends. However, the tributary valleys running across the photograph have steeper SW facing slopes (towards the camera) and gentler north-east facing slopes. This asymmetry results from microclimatic differences between opposite valley sides under periglacial conditions. TQ 042119, looking NE.

Dry valleys which dissect the chalk plateaux and dip slopes are rarely as deeply incised as those on the escarpments (photograph 66). A striking characteristic of these broader valleys, however, is their tendency to develop asymmetric cross profiles, with steep south and west facing valley sides,[32] typically 16° to 22°, and gentle east and north facing sides, usually between 5° and 9°, although valley meanders may add variability to this relationship.

Asymmetry is thought to have developed under periglacial conditions, since the long gentle north to east facing slopes are underlain by relatively thick coombe rock, up to 3 m thickness, while the steeper west to south facing slopes have bedrock close to the surface. The latter would have suffered more rapid frost shattering due to more frequent periods of thaw under higher intensity of insolation. The former would have been cold, underlain by permafrost, and suffered gradual weathering, the shattered rock moving slowly downslope in the thin active layer by solifluction. Meltwater, flowing over permafrost in the valley bottom eroded the frost shattered west and south facing valley sides preferentially, maintaining their steepness, and leading to lateral stream migration.

The periglacial origin of the dry chalkland valleys of southern England is well established. They bear witness to the effectiveness of fluvial processes in periglacial environments, where frost action and solifluction combine with spring and summer snow meltwater activity to supply abundant debris to the streams. The very high river discharges associated with the nival flood in turn lead to high sediment transport capacities, and provide sufficient stream power to enable rapid valley floor erosion to take place.

PERIGLACIAL MASS WASTING

In winter, freezing of the active layer is often associated with ice segregation within the soil. The ground surface is heaved upwards as water is drawn to the advancing freezing front and freezes on to thickening ice lenses. The result is that the ice content of the frozen ground is raised. In spring, as thaw proceeds downwards into the ground, melting ice lenses leave water-filled voids, which cannot be closed to allow the ground to resettle until the excess water has been expelled. With impermeable frozen soil below, expulsion must take place laterally, and the rate of water escape is limited by the permeability of the thawed soil. During the period of expulsion, that is, as consolidation of the soil is taking

67 Panorama of the Cheviot Hills, near Rochester, Northumberland. The rounded hill profiles result from gelifluction which has smoothed out glacial depositionary features and produced a mantle of slope deposits which thickens downslopes into the valley bottoms. Stream dissection has exposed the slope deposits (head) in channel banks. They consist of reworked till, frost weathered bedrock, or a mixture of both. NT 793078, looking NE.

place, pore water pressures are raised. This leads to a considerable loss in strength, the soil tending to behave like a viscous fluid and to flow slowly downslope under the influence of gravity. This process is called 'gelifluction' and can lead to downslope movement of the active layer at rates of a few mm to a few cm per year.[33] Where less ice forms during freezing the soil is still likely to be disturbed. Thawing leads to resettling and compaction under the influence of gravity, which produces a slight downslope displacement of the soil particles, referred to as 'frost creep'.

Measurements in a variety of periglacial environments, in both alpine areas with deep seasonal freezing, and in arctic permafrost zones, have shown gelifluction and frost creep to be the most important mechanisms of sediment removal from hillslopes. Sediment removal rates at least ten times higher than those associated with soil creep in temperate environments have been widely recorded. During Quaternary periglacial phases in Britain gelifluction was active where the bedrock was covered by sediments or a weathering mantle which contained significant proportions of sandy or silty fines. Slopes were often smoothed and rounded by gelifluction, and covered in slope deposits which thickened progressively toward the valley bottoms (photograph 67). Post-glacial stream erosion has often succeeded only in cutting a channel through the slope material (photograph 69), or in reworking some of it into river terraces.

Slope deposits resulting from periglacial mass movements, usually gelifluction and frost creep (sometimes referred to collectively as solifluction), but including mudflows, are generally called 'head'. Head is a variable material, since any sediments present on a slope, be they the product of local rock weathering, or deposited by a glacier, river, or the wind, may be subjected to gelifluction, and become incorporated into head deposits. In most cases the head is derived from local weathering of bedrock, or from a pre-existing till, and consists of a poorly sorted mixture of clay, silt, sand, gravel and cobbles. Where originating from frost weathering of bedrock, clasts are angular, and the ratio of fines to clasts as well as the texture of the fines and size of clasts depend largely on the lithology and degree of jointing of the bedrock source.

Gelifluction and frost creep results in a distinctive organisation of clasts within the resulting head. Elongate clasts tend to be orientated parallel to the direction of movement, that is, directly downslope, and to dip either parallel to the ground surface, or at a slightly lower angle, in an imbricate fashion. Although at first sight an apparently chaotic mass, therefore, head deposits are on closer inspection usually fairly well organised.

To the south of the Quaternary ice limits, in periglacial Region 1 (see the map on p. 117), head deposits are virtually universal, and derived from weathering of the underlying bedrock. In the West Country most valleys show several metres of head on their lower slopes, and the coombe rock of the chalkland dry valleys is, in fact, head. In Region 2, inside the Quaternary ice limits but beyond the Devensian glaciation, head is again widespread, and may incorporate pre-existing till. On the south coast of Gower, South Wales, for instance, frost weathering of a relic Ipswichian cliff line resulted in the accumulation of screes (photograph 68) and gelifluction affected both the weathered limestone and older till present on the plateau and in small coastal dry valleys. The result is a coastal apron of

68 *opposite* Oxwich Point, Gower. The Carboniferous Limestone fossil cliff line is frost shattered and partly buried by vegetated scree slopes. Small dry valleys cut the fossil cliff line and the gelifluction deposits that they contain merge with a coastal apron of head which extends seawards from the fossil cliff, burying a raised beach. The raised beach and fossil limestone cliffs date from the higher sea levels of the last interglacial. Scree formation and gelifluction occurred during the later Devensian glacial period, when this area lay just beyond the ice limit. Modern coastal erosion has cut a low cliff in the apron of head deposits. SS 495054, looking E.

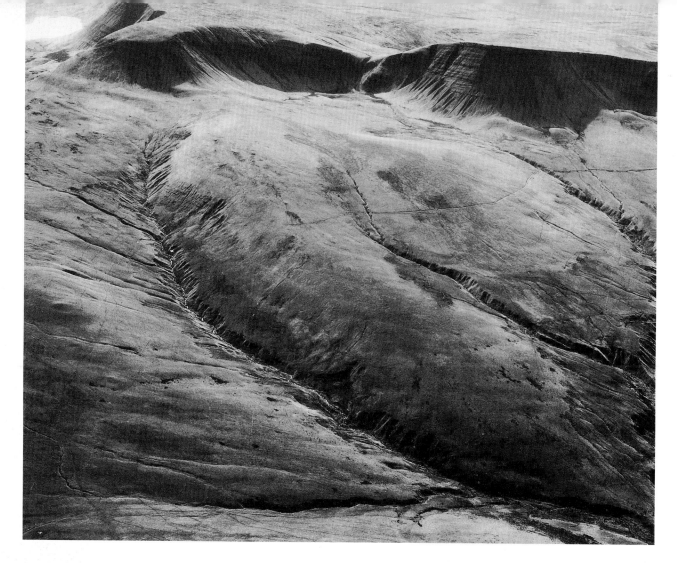

69 Afon Sychlwch, the Old Red Sandstone escarpment of the Brecon Beacons, South Wales. In the foreground gelifluction deposits are dissected by streams. Scree slopes mantle the escarpment, and protalus ramparts form an almost continuous ridge along the scree foot (see photograph 74). SN 803232, looking S.

head dominated by angular limestone fragments, but containing erratic blocks and matrix material derived from the older till.

In the Welsh uplands, the Pennines, the Cheviots, and in the Southern Uplands of Scotland, the essentially glaciated landscapes are smoothed by periglacial slope processes, and head partly derived from Devensian till and partly from bedrock weathering, is several metres thick on the lower valley side slopes (photograph 69) and in valley bottoms. The periglacial conditions under which much of this reworking of Devensian tills took place must have prevailed immediately following deglaciation, and again in the Loch Lomond Stadial, between about 11,000 BP and 10,000 BP.

In the higher mountains, such as Snowdonia, the Lake District, and especially the Scottish Highlands, the areas outside the Loch Lomond stage glaciers suffered a particularly severe periglacial climate at that time, and distinct topographic features attest to the effectiveness of gelifluction and frost creep. The most dramatic are extensive sheets of boulders covering slopes of between 5° and 35° (photograph 70) which are bounded on their lower margins by frontal banks. In the Cairngorms the large granite boulders present in these boulder sheets are associated with frontal banks up to 5 m high, but where bedrock weathers to smaller blocks, frontal banks do not exceed 3 m in height.[34] Frost creep rather than gelifluction probably caused downslope movement.

134

70 *previous page: top* Lairig Ghru, Cairngorm Mountains. Bedrock here is granite. The plateau surface in the near centre and right of the photograph consists of blockfield debris which extends as a boulder sheet downslope towards the camera. The lower limit of the boulder sheet is formed by a low bank, and the downslope mass movement of boulders occurred during the Loch Lomond Stadial. In the right background, vegetated gelifluction lobes form a series of treads and risers on the hillside. The valley sides of Lairig Ghru have numerous debris flows, with lobate frontal banks in the valley bottom. A recent ribbon-like debris flow track, with parallel ridges on either side can be seen on the valley side facing the camera, lower left of photograph. NH 984013, looking NW.

71 *previous page: bottom* Sròn na Lairige, Braeriach, Cairngorm Mountains. The summit here rises to 1,180 m. Turf-banked gelifluction lobes extend down the slope facing the camera, and occur on many of the surrounding hillslopes. On the ridge, centre left, small terracettes may be seen, with vegetated risers and bare treads. These result from the combined effects of deflation by the wind, frost heave and frost creep, and are developed on exposed ridges. Note the debris flows on the valley sides. SE 964010, looking SE.

Where weathering of bedrock during the Loch Lomond Stadial released regolith rich in fines, smaller gelifluction sheets and lobes occur, with frontal banks rarely exceeding 1.2 m high (photograph 71). The advance of these lobes and sheets buried the ground surface below, and excavation often reveals that the soil in front of the lobes may be traced upslope beneath the frontal banks. Radiocarbon dating of soil organic horizons buried by advancing gelifluction lobes indicates ages of around 5,000 years or less, indicating that frontal advance has occurred since these dates: small gelifluction lobes are at least sporadically active under the present-day maritime periglacial climate.

The high winds which characterise British mountain climates interact with frost heaving of the soil and frost creep or gelifluction to produce small scale step-like features with unvegetated deflated treads and vegetated risers (photograph 71). The risers are 0.1 to 1.2 m high and are either parallel to the contours, or trend slightly across them.[35] It appears that wind scour helps to destroy vegetation disrupted by frost action, and the lack of vegetation increases rates of frost creep and gelifluction. These small-scale features are also actively forming under the present-day mountain climate.

FROST WEATHERING

Bedrock surfaces covered with angular blocks are widely reported in the modern periglacial zone, and the obvious conclusion to be drawn is that physical weathering is active, and largely the result of freezing and thawing of water in the rocks. Despite the widespread occurrence of apparently frost-weathered regolith, the few field studies of rock temperatures which have been completed often do not show frequent cycles of freezing and thawing, especially in the high arctic. It has been suggested that hydration weathering due to wetting and drying of the rock may be of greater importance than previously thought. Chemical processes too may be significant particularly in weakening bedrock prior to frost weathering. In fact it is likely that several processes work together with freeze–thaw to produce the observed periglacial weathering of bedrock.

Laboratory studies have indicated that air temperatures must fall to below −5°C in order to promote significant rock breakdown,[36] and that the rate of rock weathering depends on the frequency of effective freeze–thaw cycles. The degree of jointing, together with pore size and frequency, all affect the susceptibility of bedrock to frost weathering, and of course water must be available. Climatic conditions favouring frost weathering are more common in alpine periglacial areas where in autumn and spring especially, daytime temperatures are frequently above zero while at night temperatures fall well below zero. As might be expected therefore, in Britain the uplands show most evidence of frost weathering, though even in the lowlands the prolonged and severe periglacial climate which prevailed beyond the Devensian ice limits also led to extensive frost weathering, producing regolith which was subsequently incorporated into the widespread head deposits.

BLOCKFIELDS

On gently sloping mountain summits and plateaux, the debris released by frost weathering may accumulate to form extensive boulder-covered areas called 'blockfields' or 'felsenmeer'. The size of boulders and the presence of finer grained material is largely controlled by bedrock lithology. In the Scottish Highlands C. K. Ballantyne[37] showed that the well-jointed rocks such as the quartzites of the North-West Highlands (e.g. Beinn Eighe, photograph 72) and the Grampians, weather to produce blockfields comprising angular boulders. Other lithologies weather to produce a greater proportion of fines. In some places till containing relatively abundant fines may also have been incorporated.

Despite the presence of finer-grained material in addition to larger rock fragments, mountain tops often present a distinctly bouldery appearance, and it is only through excavation that an abundance of fines is revealed at depth. This concentration of coarser material near the surface and fines at depth is largely due to frost sorting. This process results from ice segregation during soil freezing and is illustrated in the following diagram. Repeated freezing and thawing results in migration of clasts towards the surface while fines tend to migrate downwards.[38] Initial frost sorting of mixed grain-sized sediments therefore concentrates stones and boulders at the surface, leaving fines at depth, though as we shall see later, irregularities in the pattern of soil freezing also frequently cause lateral sorting as well, producing distinctive 'sorted patterned ground'.

72 Beinn Eighe, Ross and Cromarty. The valley sides of the glacial trough Coire Dubh Mór, on left of photograph show numerous debris flows, and a recent series of flow tracks may be seen on the steep mountain side facing the camera in the foreground. Below the cliff face in the middle right of the photograph are screes with well developed fall-sorting of clasts, the boulders having rolled to the bottom of the scree and the smaller clasts having lodged higher up. Note the late snow patches occupying small nivation hollows. NG 965595, looking W.

Process of frost sorting according to the 'frost pull' hypothesis. 1–3, freezing stage, accompanied by ice segregation and frost heave. 4–6, thawing stage, accompanied by water expulsion and consolidation as voids left by melting ice lenses are closed. At stage 2 the boulder is frozen onto the soil above. As ice segregation at the freezing plane causes frost heave, the boulder is pulled upwards. At stage 5 the soil consolidates and settles as it thaws, while the boulder remains supported by the still-frozen soil below.

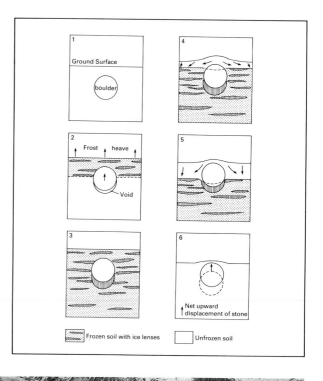

73 An Garbh Choire, Cairngorms, with the ridge of Braeriach (1,296 m) on the left of the photograph. The stony debris-covered plateau surface has been subjected to frost weathering. Subsequent wind action has removed the finer grained material, leaving a residual stony lag. Note the well-developed linear nivation hollow occupied by a late snow patch, lower right of photograph. When the snow has melted a distinct bench-like 'altiplanation terrace' will be left on the hillside marking the area where nivation has been active. Stony gelifluction sheets with sharply defined frontal banks can be seen on the plateau, middle left of the photograph. NN 938980, looking NE.

On certain lithologies in the Scottish Highlands the fines released by weathering are sandy and highly susceptible to deflation by the wind. The Torridonian Sandstone mountains of the North-West Highlands and the granite mountains of the Cairngorms fall into this category,[39] and deflation of fines in these areas often leaves a lag of gravel covering plateaux surfaces (photograph 73). On lee slopes sand is redeposited in sheets which are generally well vegetated.

Blockfields are extensively developed in British mountain regions where the surface gradient is low. At lower altitudes, however, such as in the Pennines and Brecon Beacons, fossil blockfield surfaces are frequently vegetated or covered in peat. Even on the highest mountains, despite their unvegetated surfaces, blockfields are today not active, but rather owe their formation to intense frost weathering during the Loch Lomond Stadial.

SCREE

Glacial erosion of upland valleys has left them with steep, often almost vertical sides. The newly exposed bedrock surfaces were highly unstable immediately following deglaciation and rock fall activity was almost certainly high. During the subsequent Loch Lomond Stadial, frost weathering of the valley sides released abundant angular rock debris ranging in size from small fragments of rock to large boulders, which accumulated against the valley sides as screes or talus slopes.

In photograph 74 the Devonian Old Red Sandstone escarpment of the Brecon Beacons, S. Wales, is partly covered by scree which extends into the small corrie

74 Benan Sir Gaer, Old Red Sandstone escarpment, Brecon Beacons, South Wales. Screes mantle the lower parts of the escarpment, showing typical straight, upper and middle sections and a marked basal concavity. Post-glacial erosion has dissected the screes, which are no longer actively accumulating. At the base of the scree, in the centre of the photograph is a protalus rampart picked out by dark-coloured bracken along its crest. This ridge can in fact be traced along much of the scree in the left foreground, and further east in photograph 69. SN 812219, looking SW.

139

basin in the centre of the photograph. The corrie was reoccupied by ice during the Loch Lomond Stadial so that its screes must date from the later part of that cold period, following disappearance of the glacier. It is noticeable that the screes within the corrie are considerably smaller than those outside.

Rockfall activity from a cliff face produces screes inclined at angles close to the angle of repose of the rockfall debris. This is the maximum slope at which the debris is stable, usually between 35° and 45°.[40] Accumulation of debris may increase the scree slope gradient but when it reaches its critical angle of stability further falling debris will slide over the surface. In fact, since the debris arrives with a certain momentum, it will tend to roll and slide over a slope slightly below the angle of repose[41] so that screes are usually inclined at this slightly lower angle. Where debris is supplied preferentially from gullies and chutes in the cliff face, fan-shaped scree units develop and merge to form a more complex scree (photograph 75).

Unless confined by a valley bottom, or subject to basal erosion, screes show straight upper and middle slopes with angles close to the angle of repose of the

75 Coire Garbhlach, Cairngorms. Cone-shaped screes have coalesced along the valley sides. These screes are traversed by numerous debris flows. NN 885945, looking E.

scree material, and a marked concavity at the base (photograph 74). Associated with this concavity is an increase in size of clasts downslope, so that the foot of the scree may consist of an apron of large boulders (photograph 72). To understand this 'fall sorting' of scree we must consider the roughness of the scree surface in relation to the size of falling debris. Rock fragments falling on to the scree are more likely to lodge quickly rather than roll or slide over the slope if they are small relative to the surface roughness, since they rapidly encounter a hollow into which they can lodge. Larger blocks however, will slide further since the surface will be relatively smooth compared to their diameters. In consequence, smaller particles tend to accumulate near the top of the scree while larger particles slide and roll further down the scree. The greater momentum of the large clasts will also tend to carry them further downslope.

Most screes in Britain are not active today. In the Scottish Highlands for instance, scree volumes have been used to estimate rates of rockwall retreat during the Loch Lomond Stadial, and suggest average values of around 2 mm per year, which contrasts markedly with estimates of present day rates of retreat of around 0.015 mm per year.[42]

DEBRIS FLOWS

Related to screes, and indeed often affecting scree slopes, are the deposits left by debris flows. Debris flows consist of rapidly flowing water-charged masses of rock fragments, and the main contrast between debris flow deposits and fluvial deposits is that the former remain unsorted during flow and deposition, while in the latter bedload and suspended load are transported and deposited as distinct units. Debris flows occur mainly on steep hillsides as a result of rapid infiltration of water into regolith during high intensity rainstorms, or rapid snow melt. The debris becomes unstable as pore water pressures rise, and is mobilised as a rapidly moving slurry. Loss of water along the sides of the flow leads to sediment deposition which builds up levees on either side of the flow track (photographs 70, 71 and 72). At the base of the slope where the debris flow terminates, these levees broaden and merge to form lobate toes.

Debris flows occur frequently in the Scottish Highlands today, with at least 71 such flows recorded in the steep-sided glacial trough of Lairig Ghru, Cairngorms (photograph 70) between 1970 and 1980.[43] Many flow tracks are repeatedly reoccupied, and there has been a marked increase in debris flow activity in the Highlands of Scotland during the nineteenth and twentieth centuries, probably as a result of vegetation changes due to burning and overgrazing.

PROTALUS RAMPARTS

Examination of photograph 74 reveals a distinct ridge at the base of the scree, running parallel to the contours. Although less clearly defined in places, this ridge is more or less continuous along the bottom of the scree on the eastern (left hand) side of the photograph. The ridge is composed of angular clasts and provides an excellent example of a late-glacial protalus rampart. During the cold periods immediately following deglaciation and again during the Loch Lomond

Formation of protalus rampart.

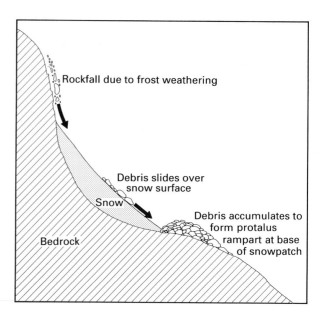

Stadial, large snow patches developed against this escarpment as snow drifted over the crest. Since the scarp is north-facing, exposure to the sun was minimal, and the snow patches probably survived throughout the summer. Debris released by frost weathering fell on to the snow surface, and since this was steep, slid to the base, and accumulated to form a ridge of talus material, that is, a protalus rampart. The eastward extension of this ridge may be clearly seen in photograph 69. Protalus ramparts consisting mainly of boulders and pebbles have been described in North Wales, the Lake District, and the Scottish Highlands.[44]

The very large ridge at Baosbheinn, Wester Ross, illustrated in photograph 45 in chapter 3, was considered to be a very large protalus rampart, but it may have resulted from massive landsliding which over-rode a large snowbed, leaving a ridge of enormous boulders, rather than the gradual accumulation of rockfall debris characteristic of true protalus ramparts.[45]

NIVATION

Late lying snow patches on bedrock rather than scree are frequently associated with distinct hollows in the hillside, suggesting that weathering is enhanced by the presence of snow. In the Rocky Mountains of Colorado, field studies have shown that around the edge of a snow patch, although not underneath it, the frequency of freezing and thawing is increased, and meltwater provides the moisture necessary for frost weathering to occur.[46] Since ablation through the summer gradually reduces the size of the snow patch, a broad area of ground in the vicinity of the patch may suffer enhanced weathering. C. E. Thorn also measured accelerated slope wash and chemical weathering associated with snow patches, and many authors have reported increased rates of gelifluction immediately below patches.[47] Hence small irregularities in the ground surface which may encourage snow to accumulate, become enlarged into 'nivation hollows', often with gelifluction lobes immediately below. The term 'nivation' describes the range of processes responsible for accelerated erosion in and around snow patches.

In the foreground of photograph 73 a linear snow patch may be seen occupying a distinct nivation hollow. Such late lying snow is today relatively ineffective, though nivation features still occupied by late snow patches have been noted in the Lake District and in the Scottish Highlands.[48] Under the severe periglacial climates prevailing in the past, however, nivation has in places produced broad step-like benches cut into hillsides. Such features are called 'altiplanation terraces' and have been described from Dartmoor and the Brecon Beacons.

TORS

Tors consist of upstanding residual masses of bedrock rising conspicuously above the surrounding ground surface. They result from differential weathering of bedrock so that the more resistant areas survive longer and remain higher than the intervening less resistant areas. Two distinct origins have been proposed for tors in Britain, the so-called two stage theory of D. L. Linton and the one stage theory of Palmer and Radley.[49]

The Linton model was developed with reference to the granite tors of Dartmoor and the Cornish moors, where Tertiary deep chemical weathering was considered to have occurred, particularly where the rock was highly jointed. Subsequent Quaternary erosion left the less well jointed areas, where intact granite survived, as hills with summit tors. These tors consist of rounded 'core stone' rock outcrops surrounded by granite boulder fields referred to as 'clitter'.

In contrast, Palmer and Radley described Millstone grit tors in the Pennines where deep chemical weathering has not occurred. They concluded, on the basis of the angularity of the tors and their surrounding clitter, that frost weathering under Quaternary periglacial climates combined with gelifluction of the

76 The Devil's Chair tor, Stiperstones, Shropshire. The tor is formed of steeply dipping quartzite, and is surrounded by angular boulders released by frost weathering.

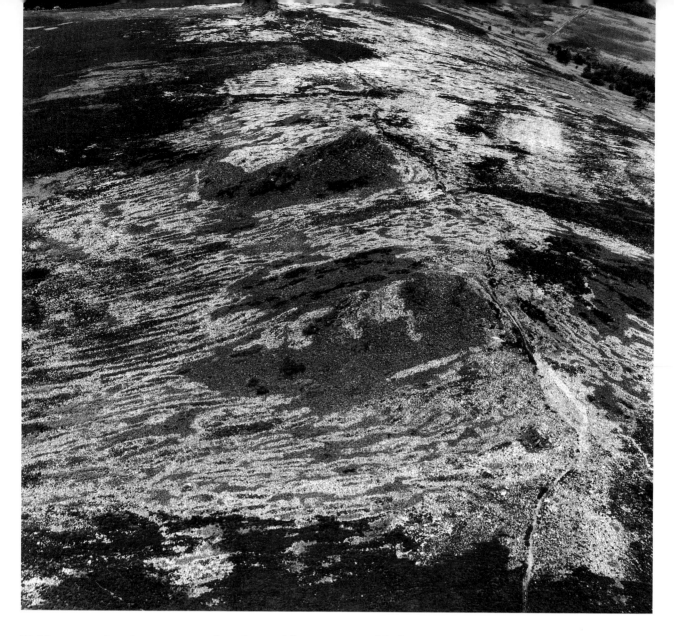

77 Stiperstones, Shropshire. This quartzite ridge has a maximum altitude of 525 m, and prominent tors outcrop along it. The largest, the Devil's Chair, can be seen in the background, and the tor in the middle distance is called Manstone Rock. Large-scale sorted stripes are clearly defined on the bouldery slopes below the tors. In the flat area of low gradient immediately in front of Manstone Rock is an area of irregular sorted polygons which become elongated and give way to stripes as the gradient increases on either side of the ridge crest. SJ 361984, looking N.

weathered material were responsible for exposing the tors. It is, however, unlikely that a single weathering process is responsible for all tor formation, since in Victoria Land, Antarctica, for example, both angular frost riven tors and rounded core stone type tors have developed under an extreme polar desert climate.[50]

A superb example of tors occurs in Shropshire on the Stiperstones ridge (photographs 76, 77). The quartzite bedrock dips steeply to the north-west and tors rise abruptly above the smooth ridge crest. The largest tor, the Devil's Chair, stands about 20 m above the adjacent ground surface.[51] The slopes of the ridge are covered in angular blocks derived from the tors. Sorting of the regolith into bouldery stripes aligned parallel to the slopes is widespread (photograph 77), and where slopes are less than 7.5° polygonal networks with diameters around 7 m to 9 m are present. On slopes of between 6° and 16° these polygons are markedly elongate downslope, and merge into stripes.

The periglacial origin of these sorted stripes and polygons is beyond dispute. Frost weathering under severe periglacial conditions was responsible for exposing the tors under severe periglacial conditions during the Devensian ice maximum

144

period between about 25,000 years and 15,000 years ago. At this time the Stiperstones ridge lay immediately adjacent to the ice sheet margin (see the map on p. 117), and only escaped being engulfed by ice on account of its superior altitude.

SORTED PATTERNED GROUND

The sorted stripes and polygons of the Stiperstones (photograph 77) are picked out by vegetation growth in the centres of polygons and between the stone stripes. Growth of this vegetation is promoted by concentrations of fines in these areas. The frost shattered debris has therefore been subjected to lateral sorting to produce alternating stripes of boulders and fines on steeper ground, and polygonal networks of boulders with fines concentrated in the centres in flatter areas. Similar sorted stripes occur on Dartmoor, in North Wales, the Lake District and the Scottish Highlands. In the Rhinog Mountains of North Wales sorted stripes with spacing 5 m to 8 m have been recorded, and as in the Stiperstones example, boulders in the stripes showed a marked downslope orientation.[52]

Similarly, in the Highlands of Scotland large sorted circles and stripes occur in blockfields where adequate fine-grained material is present. On Rhum, in the Inner Hebrides, for instance, circle diameters are between 2 m and 3 m, while elsewhere diameters up to 4.5 m are reported. As the slope angle increases the circles and polygons become elongated downslope, and above approximately 7.5° they merge into stripes.[53] Much smaller sorted circles and stripes with circle diameters 15 cm to 50 cm also occur on many higher mountains, such as Snowdonia, the Lake District and the Scottish Highlands.[54] The internal structure of these features is identical with the larger forms but on a smaller scale, as shown in the diagram. The larger features are entirely fossil under the present climate, but the smaller forms are actively forming today.

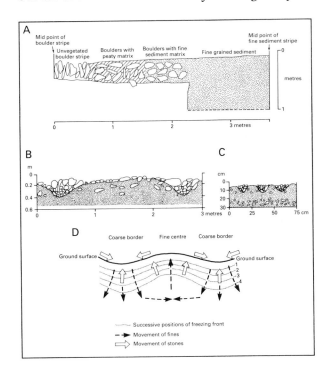

Internal structure of sorted patterned ground.

(A) Large-scale sorted stripes, Rhinog Mountains, North Wales.

(B) Large-scale sorted circle, Sron an t-Saighdeir, Rhum.

(C) Small-scale sorted stripes, Lake District.

(D) Postulated pattern of sediment movement in sorted circles.

The mechanism of upward migration of clasts due to freezing and thawing of the ground has been described above. In order to produce the observed sorted patterns, however, some mechanism of lateral sorting must exist. One possibility is that an irregular sinuous freezing plane penetrates the ground, due to its natural thermal variability. This would tend to cause clasts to migrate towards areas of more rapid freezing and fines to concentrate where freezing rates are less. The role of permafrost has not been clearly defined, but it has been suggested that sorted circles of diameter greater than 2 m are always formed in association with permafrost.[55] A recent model to explain the regularity of sorted circles and polygons invokes thermally driven Rayleigh convection of water in the active layer during thaw,[56] and mass displacement of waterlogged fine sediments due to density differences in the active layer may also be important in some cases. The transition to sorted stripes on steeper ground is clearly related to gelifluction, the fine grained stripes moving downslope more rapidly than the coarser stripes.

Whatever the precise mechanisms of formation however, by analogy with modern periglacial areas, the large scale sorted patterns of the British uplands date to a period of severe periglacial climate when continuous permafrost was present. The depth of sorting may well relate to the thickness of the active layer. Inside the Devensian ice limits the most likely such period was the Loch Lomond Stadial. The small scale features however, may be currently active, and result from periodic freezing of the ground to a few tens of cm at most.

CONCLUSION

The wide range of periglacial landforms still visible in Britain today has been illustrated in this chapter. Probably the most important features not shown, however, are the periglacial river deposits occupying such major river valleys as the Severn and Thames in southern England. These form terraces along the river courses, and where exposed in gravel pits show sedimentological characteristics of braided stream deposits. Frost weathering, gelifluction and the nival spring flood combined to produce high concentrations of coarse sediment in these periglacial rivers. It is perhaps a salutory reminder of the importance of Quaternary cold climates in the evolution of British landscapes that approximately 95 per cent of the river deposits present in the Thames Valley were laid down under periglacial conditions.

GUIDE TO FURTHER READING

H. M. French, *The Periglacial Environment*, London, 1976.

A. L. Washburn, *Geocryology*, London, 1979.

J. Boardman (ed.), *Periglacial Processes and Landforms in Britain and Ireland*, Cambridge, 1987.

5   Fluvial landforms

INTRODUCTION

Rivers and streams, both past and present, have played a major role in fashioning the landscapes of Britain. Even in heavily glaciated areas, glacial erosion tended to be concentrated in valleys created by preglacial rivers and post-glacial landscape development of glaciated uplands and lowlands alike has been largely the result of fluvial action. Rivers tend to play a dual geomorphological role: an 'active' erosional role in valley formation and development via downcutting, lateral erosion and headward growth and their influence on valley-side slope angles and processes; and a 'passive' role in transporting, sometimes with complex and long-term lags and stages involved, the sediment and dissolved material supplied to them by hillslope denudational processes. In carrying out these roles, rivers are strongly influenced by other factors, notably lithology, structure, tectonic history, current climate and climatic history, and increasingly human activity.

Fluvial geomorphology in Britain has altered and developed considerably over the last 30 years. Prior to the 1960s, fluvial geomorphology as a separate field of study did not exist. Although this in part reflected a lack of river-flow data in Britain with which to pursue such studies, it was mainly due to the dominance of geology-related long-term landscape evolution studies within geomorphology at that time.[1] Studies of rivers were descriptive and qualitative in nature and mainly focussed on such aspects as the long-term evolution of river systems and the interpretation of river long profiles. These studies relied on logical deduction and argument in explaining successive stages in landscape development, but processes and mechanisms invoked to explain landscape change were largely inspired guesswork and had very little empirical basis.

The later 1960s saw the growth of fluvial geomorphology into one of the main branches of geomorphology with the advent of quantitative drainage basin analysis and river process studies, as new ideas and approaches from the United States[2] were imported and developed by R. J. Chorley at Cambridge and systems geomorphology[3] provided a new framework for fluvial studies. This new phase of fluvial geomorphology was hydrology-based and aided by a parallel expansion in the number of river-flow gauging stations.[4] The drainage basin became the fundamental areal unit of study, as it was viewed as an integrated sediment-water-landscape system. Relationships were explored between discharge variables and river channel cross-sectional form and pattern, and knowledge of current processes, such as the suspended and dissolved loads of rivers, river bank erosion, meander migration and floodplain dynamics, was greatly expanded.[5]

Two trends can be observed in contemporary fluvial studies, reflecting both the results and shortcomings of the process geomorphology phase. First there has been a re-expansion of time-scales. This is because many fluvial features cannot be explained adequately in terms of current processes alone, but are the outcome of an overlapping series of adjustments to a Pleistocene legacy, post-glacial climatic changes and the history of human activities particularly over the last two millenia.[6] Indeed, it is the impact of these activities on rivers which has constituted the second major trend in British fluvial geomorphology. These changes in fluvial geomorphology in Britain and the use of aerial photography in fluvial studies form the two central themes of this chapter.

AERIAL PHOTOGRAPHS IN FLUVIAL GEOMORPHOLOGY

Aerial photographs have proved an invaluable tool to the fluvial geomorphologist in several ways and have contributed significantly to many of the advances made in fluvial studies since 1960. Vertical air photography has been widely used in morphometric studies as, subject to scale and quality constraints, it can be used to derive highly detailed and accurate maps of specific fluvial features, such as river channel networks, meanders, floodplain morphometry, channel width, river banks, and (depending on flow conditions at the time of photography) channel features such as gravel bars, shoals and riffles. The detail is usually far better than on Ordnance Survey maps because only specific features of interest to the compiler are being mapped. Repeated vertical photography allows changes in fluvial features through time to be assessed and in many studies the timescale has been extended by comparing aerial photographs with early Ordnance Survey and other historical maps. A number of examples of this type of study are referred to below. The utility of aerial photographs for historical studies has of course

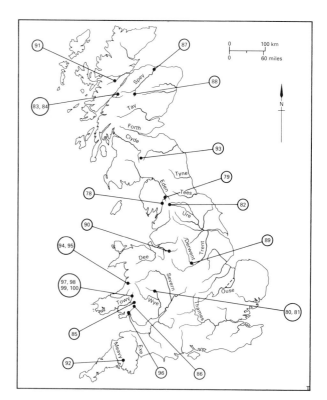

Location of the aerial
photograph sites.

been increasing yearly as the timespan covered by photographs, now over 40 years, has extended. Important constraints, however, are the tendency for vegetation to obscure stream channels and floodplain features and the difficulties of accurate contour mapping of the typically low relief of floodplain/channel areas. Also, oblique photographs are difficult to use for mapping purposes except in a very qualitative way, but they can be valuable in identifying and dating channel changes.

Apart from their use in mapping fluvial landforms and helping to assess and date changes in river channels, aerial photographs can show up (via different crop or grass marks) fluvial features not visible on the ground, such as old river courses or different types of floodplain deposit. In this way, even provisional analysis and interpretation of oblique or vertical air photographs at the reconnaissance stage of investigation can lead to the identification of past fluvial landforms and aid the planning of more detailed field investigation of the features. Furthermore, the very different and wider perspectives offered by aerial photographs, particularly obliques, can help geomorphologists to perceive fluvial landscapes as assemblages of inter-related features rather than as individual landforms. Finally, obliques afford an invaluable teaching and research visual aid in helping overcome the conceptual gap between field observation and textbook maps and diagrams. This chapter endeavours to demonstrate some of these uses to which fluvial geomorphologists have put aerial photography.

PRE-HOLOCENE DRAINAGE DEVELOPMENT AND CHANGE: SOME EXAMPLES

Until the 1960s long-term landscape development was the main focus of study of geomorphologists in Britain. As rivers were the main agents of erosion and hence landscape evolution in Britain, investigations of drainage evolution formed an important component of these 'denudation chronology' studies. Furthermore, in these landscape and drainage evolution studies, 'historical geomorphologists' drew heavily on fluvial evidence in the present landscape. The sites and altitudes of abandoned cols formerly occupied by captured or diverted rivers, irregularities (knick points) in river long profiles, river terrace remnants, elbows of river capture, channels too small for their valleys and the lithology and calibre of fluvial sediments are but a few of the lines of fluvial evidence which have been employed. Detailed consideration of the principles controlling long-term drainage evolution and how it has been influenced in Britain by such factors as lithology, structure, tectonic history, changes in 'base' levels (both sea-level and local base levels) and changes in climate is beyond the scope of this chapter. The examples given are intended to demonstrate the nature of, and some of the limitations, problems and uncertainties associated with, this now sadly rather neglected part of British geomorphology. For a modern example of historical geomorphology, the reader is referred to D. K. C. Jones' comprehensive and fascinating account of the geomorphological history of South-East and Southern England from the Tertiary to the present day.[7]

In Northern England, King[8] has examined drainage evolution in the Howgill Fells area, involving the headwaters of the Lune, Eden and Ure rivers. The diagram

Long-term drainage
development in the Howgill
Fells area.

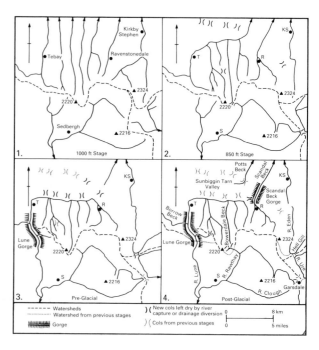

above summarises stages in the development of drainage of the area. The Howgill Fells upland is an eastward extension of the Lake District and is composed largely of Silurian slates with overlying Carboniferous rocks outcropping to the north and south. The most recent major uplift occurred along an east–west axis in the Tertiary, forming a monoclinal block with a steep southern slope and a gentle northern one. The identification of a series of erosion surface remnants at altitudes ranging from 610 m down to 90 m is interpreted to mean that uplift and the associated fall in base level (the sea) were episodic and continued at least into the late Tertiary and perhaps into the Pleistocene. The first stage represents a simple, structurally controlled drainage pattern, which was initiated as the east–west Howgill Fells uplift axis emerged. It consisted of a series of low gradient north-flowing streams draining the gentle northern slope and steep, south-flowing streams draining the steeper southern slope of the Howgill Fells uplift. The fact that many of the courses of these streams are discordant to the structure of the Silurian rocks is interpreted as suggesting that they may have been initiated on a thin (now eroded) layer of Mesozoic rocks and later superimposed onto Carboniferous and Silurian rocks during subsequent denudation. The watershed between the Lune and the Eden at this time lay well north of its current position.

By Stage 2, the development of an east-flowing subsequent along the weaker beds of the Ashfell sandstone within the lower Carboniferous (and flowing into what is now Potts Beck and hence northwards to the Eden) had captured the southern headwater sections of the north-flowing consequents and had left a series of dry cols at 300 m altitude in the Carboniferous limestone scarp.

Stage 3 involved the headward growth of the Lune northwards to breach the east–west watershed, capture and reverse the drainage of the former north-flowing stream, and led to downcutting of the reversed drainage section south of Tebay to form the Lune Gorge (photograph 78). The line of reasoning used to explain why the Lune grew headward at the expense of northward flowing

drainage involves the 'law of unequal slopes', which states that, other things being equal, the stream with the steepest gradient (in this case the south-flowing Lune) will incise and grow headward more rapidly and hence extend its drainage basin at the expense of adjacent streams of lower gradient (in this case the north-flowing streams). At some stage after this, once the Lune exposed the weak basal conglomerate at the junction of the Carboniferous and Devonian just north of Tebay, a subsequent strike stream developed rapidly eastwards along the outcrop towards Ravenstonedale. This resulted in (1) the progressive capture of the upper reaches of the original north-flowing consequents, including the Scandal Beck (photograph 79), and (2) a second line of beheaded streams and dry cols to the north and a misfit Potts Beck. The Lune system had thus extended its drainage basin northwards around 8 km at the expense of the Eden by the end of this third stage.

78 The Lune Gorge was formed when the strongly flowing Lune, running down the steeper southern slope of the Howgill Fells Uplift, cut headwards to breach the divide and capture former north-draining streams including the Borrow Beck (the confluence visible centre left). The breach may have been deepened by later glacial action. NY 613005, looking N.

79 Scandal Beck Gorge with Ravenstonedale village and the Howgill Fells in the background. The 75 m deep gorge was formed when meltwater, impounded in the upper Lune Valley as ice retreated westwards, escaped northwards using a col left dry by the capture of the Scandal Beck by the Lune in preglacial time. The Scandal Beck was thus re-diverted to the north-flowing Eden system. The Gorge at NY 730080, looking SSE.

Stage 4 involved glacial action. As ice retreated westwards towards the Lake District along the upper Lune Valley, meltwater became impounded in the Raven-stonedale area between the retreating ice and the high ground of Ash Fell Edge (NY 735048). The meltwater escaped northwards using the col left dry at 270m altitude by the capture of the upper reaches of Scandal Beck by the Lune in Stage 3. These meltwaters were able to incise deeply to form the 75 m deep Scandal Beck Gorge (photograph 79) and divert the entire Scandal Beck back to the Eden drainage system, leaving a new, low dry col between the Scandal Beck and the Lune (see the diagram on p. 150).

The neatness and elegance of denudation chronology accounts such as the above tend to disguise the problems and uncertainties involved. Three types of

problem in particular tend to occur. The first concerns the nature and validity of some of the process mechanisms involved in such studies, such as river capture and the law of unequal slopes in the example above. The second problem is that hypotheses are too often based on lack of evidence against rather than positive evidence for a postulated drainage change or sequence of events; often alternative hypotheses fit the few available facts equally well. The third problem concerns the dating of events; events are placed in chronological order, but the absolute timing of events is often very vague.

Exactly how watersheds are breached and river captures and drainage reversals effected by headward-cutting streams is unclear, particularly as the area generating the overland flow or seepage responsible for headward growth and incision would become progressively reduced as the divide between competing streams is encroached. Equally, much is assumed in the law of unequal slopes about a simple relationship between channel gradient and rate of downcutting. However, there is little empirical evidence to prove (or disprove) both these generalisations. Furthermore, recent process geomorphology, with its accent on rapid, short-term rather than slow, long-term process studies, has not really tackled these questions and little is known about rates of downcutting and headward growth especially on hard rock as opposed to unconsolidated material. River capture has thus usually been deduced from evidence of pre-capture drainage links. The evidence ranges from inconclusive forms of evidence (such as sharp changes in river direction and misfit streams) to more conclusive evidence yielded by analysis of the long profiles and sedimentological character of valley floor deposits and river terraces in the upstream 'captured' and downstream 'beheaded' river sections. Dating is a notoriously weak point of denudation chronology and many old hypotheses have had to be revised with the advent of radiocarbon and isotope dating techniques in the 1960s. Thus King, although mapping the Lune Gorge (photograph 78) and the extension of the Lune to Ravenstonedale as 'pre-glacial' (see the diagram on p. 150), also stated that the watershed breach must have occurred in the early Pleistocene and that it was possible that ice may have helped to complete the breach by glacial diffluence.

It was because of the scepticism and uncertainty surrounding denudation chronology and the inapplicability of process studies and new quantitative techniques to the solution of denudation chronology problems that this branch of geomorphology waned in importance. Some geomorphologists, however, transferred their denudation chronology skills to the study of more recent landscape development in the Quaternary, where glacial and interglacial deposits provided the process and dating evidence so lacking in many preglacial landscape problems. Furthermore, the glacial period witnessed numerous drainage changes in Britain. Drainage diversions, both temporary and permanent, resulted from watershed breaching by glaciers, blocking of river courses by ice advance, overflows (and channel cutting) of ice-dammed lakes, meltwater channel erosion, and blockage by glacial, fluvioglacial and lacustrine deposits.

A striking example of the scale and complexity of such changes is given by the diversions of the River Lugg and River Teme in Herefordshire, as shown on the following map.[9] In pre-Devensian times, the Lugg was a west-bank tributary of the then south-flowing Teme and it reached the Teme via a more southerly

80 Glacially diverted course of the River Lugg, near Aymestrey, Herefordshire, looking west. The eastward advance of the Wye Glacier to the South blocked the Lugg and forced it ENE between the two wooded hills in the foreground, Mere Hill, left (SO 408654) and Sned Wood, right (SO 403661). Also visible in the middle-top left is part of the Covenhope Gap, which the diverted Lugg first used before it too was blocked by ice to the south.

Drainage changes in the Lugg and Teme resulting from glaciation.

route through Combe Moor and Shobdon than the current tortuous, gorge route from Kinsham to Aymestrey (see the map and photograph 80). In the Devensian, the eastward advance of the Wye Glacier blocked the preglacial Lugg at Combe Moor and the river cut a new course to the north-east before turning south-east through the east-west ridge via the Covenhope Gap. As the Wye Glacier advanced eastward, however, this route too became blocked and the Lugg cut a third course via a gorge (photograph 80) to join the preglacial Teme at Aymestrey. With further ice advance eastwards and northwards, the combined Lugg and Teme at Aymestrey was blocked leading to the formation of glacial Lake Wigmore, which overflowed north-eastwards forming Downton Gorge. For a time the diverted Teme flowed southwards through Orleton and Leominster until the Wye Glacier in turn dammed the Teme at Orleton to form glacial Lake Woofferton. The lake overflowed eastwards, reversing the drainage of a previously west-flowing stream from Tetbury Wells and downcutting to form Shelsey Gorge. During deglaciation, the Irish Sea ice from the north decayed faster than the Wye Glacier and consolidated these new east–west links of the Teme, as the southerly routes remained blocked, first by ice and then by glacial till and lacustrine deposits. The Lugg readopted the most easterly of its former courses (photograph 80), as this was the first to be free of ice and lacustrine and glacial outwash deposits blocked

81 The 'misfit' River Lugg south-east of Mortimer's Cross, where it is an inactive, small, meandering river in a wide floodplain formerly occupied by a combined Lugg-Teme river before the Teme was diverted eastwards in glacial times. Note the minor river terraces picked out by the low-angle late afternoon sun. River Lugg at SO 436627, looking NE.

the northern exit into the former Lake Wigmore area and to the Teme. The old courses via Covenhope and Combe Moor had been plugged with considerable depths of glacial till by the Wye Glacier and hence were not re-adopted by the Lugg after further glacial retreat. Indeed the Covenhope gap is currently occupied by a north-flowing tributary of the Lugg. These drainage changes have had profound implications for the fluvial landforms and river character of the post-glacial Lugg and Teme. The Teme's post-glacial course is now irregular and characterised by a series of gorge and basin sections. The Lugg downstream of Aymestrey is now a markedly 'underfit' river, occupying a large floodplain related to the former Teme-Lugg-Onny river. Consequently the meanders and channel are very small compared with the size of the inherited valley (photograph 81).

FLUVIAL LANDFORMS OF DEGLACIATION AND POST-GLACIAL ADJUSTMENT

Major fluvial activity occurred during deglaciation and early post-glacial times as streams and rivers, their discharges inflated by meltwater and enhanced runoff from unvegetated slopes, encountered the glacially disrupted landscape. The scope for sediment transport, floodplain sedimentation, vertical and lateral erosion and terrace development by post-glacial rivers was immense since highly erodible till, outwash, lacustrine and periglacial deposits covered much of the landscape and in particular were concentrated in the preglacial valleys of rivers. Rivers also had to adjust to stream courses and gradients usually both irregular and very different from those of preglacial times. Some of the landforms and river features which resulted are now discussed with reference to aerial photographs.

The post-glacial long profile of the Hearne Beck was altered significantly as a result of its diversion by a drift tail deposited by the eastward moving Wensleydale glacier across its junction with the Ure.[10] The channel slope was reduced as it flowed behind the drift tail but a waterfall, Hardrow Scar, resulted as the stream rounded the tail and descended sharply to the River Ure (photograph 82 and diagram, p. 157). Post-glacial erosion during the past 10,000 years by Hearne Beck has resulted in a 1 km retreat of the waterfall to form a gorge section downstream of the current position of Hardrow Scar. The rapid rate of waterfall retreat has been aided by lithology and structure, as soft shale underlies hard, horizontally bedded sandstone and limestone. Similar diversions, waterfalls and gorges were produced in the same way at other locations in Wensleydale.

Many river terraces in the British landscape date from deglaciation and early post-glacial times. Defined as lateral benches between a river channel and its valley sides,[11] river terraces are produced when a river incises a previous floodplain (the terrace) as a result of a change in base level, a change in discharge-sediment conditions in the river catchment, or a combination of the two. Deglaciation and post-glacial conditions were ideal for terrace formation with massive sediment availability for floodplain construction, river long profiles completely unadjusted to post-glacial conditions, variable discharge regimes resulting from climatic change and changing meltwater contributions, and ice-dammed lakes, sea level

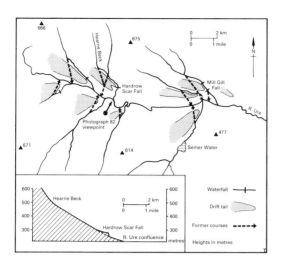

Diversion of Hearne Beck and adjacent streams in Wensleydale by downvalley drift tails at tributary junctions.

82 Hardrow Force, a waterfall on Hearne Beck in Wensleydale. The lower course and confluence of the Hearne Beck with the Ure were diverted eastwards (downvalley) by a glacial drift tail deposited by the eastward moving Wensleydale glacier. The waterfall has retreated about 1 km during post-glacial time. SD 869917, looking NE.

83 River terraces of the lower River Roy at Bohenie on the right bank and Bohuntine on the left. These terraces graded to successively lower levels of a proglacial lake in lower Glen Spean during deglaciation. Also visible is the deep drift in the valley, which marked the limit of the Loch Lomond Readvance (Stadial) up Glen Roy and which caused a lake to persist behind it into post-glacial time. The early post-glacial terraces of lower Glen Roy only extend as far as this drift dam, showing the incision of this dam did not occur until later in post-glacial time. NN 294829 (right bank) and NN 291834 (left), looking N.

recovery and fluvial incision providing numerous instances of changing local or regional base levels.

The terraces of Glen Roy are amongst the most dramatic and best documented in Britain (photographs 83, 84a,b and the map on p. 160). However, although all were produced by the River Roy during deglaciation and early post-glacial time, the higher terraces of the lower Roy (photograph 83) are neither contemporaneous with nor connected to the higher terraces of the upper Roy (photograph 84), as the base level controls were very different. J. B. Sissons, who investigated the lower Glen Roy and Glen Spean terraces, identified and mapped over 20 different terraces, of which he was able to trace and reconstruct the long profiles and down-valley extent of 18.[12] These he numbered according to their altitude, with No. 1 the highest (and oldest) and No. 18 the lowest (and youngest). Four of these (2, 15/16, 17 and 18) are prominent in lower Glen Roy (photograph 83 and the accompanying map) but none extend into upper Glen Roy.

(a)

84a, b Major river terrace
sequences of Upper Glen Roy.
The highest lies over 30 m
above the present River Roy,
preserved (a) above and below
the confluence with the River
Turret (foreground) close to
Brae Roy Lodge and at the
confluence with the Allt na
Rheinich (near background),
looking SE; and (b) further
downstream at the confluence
with the Allt Brunachain,
looking SE. The high terraces
appear to grade to levels of the
early post-glacial lake formed
behind the drift dam seen in
photograph 83; the lower
terraces grade to successive
stages in the later incision of
the drift dam by the River Roy.
(Note also the 'Parallel Roads'
on the valley-side formed by
shoreline erosion at different
glacial lake levels at the height
of the Loch Lomond Readvance.)
A. Looking SE from NN 343926
B. Looking SE from NN
318897.

(b)

River Terraces of Glen Roy
resulting from deglaciation in
early post-glacial times.

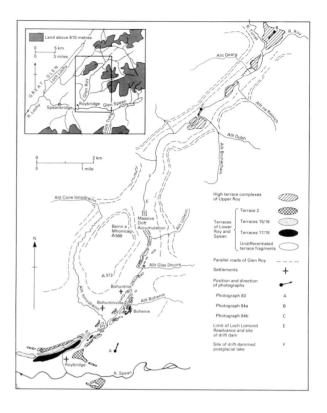

The terraces and their varying altitudes and spatial extent are intricately related to, and explicable only in terms of, the exact sequence of events during deglaciation and immediate post-glacial time. The main part of the Devensian glaciation in Scotland ended around 13,000 BP and was followed by a Loch Lomond Interstadial with ice-free conditions in the Scottish Highlands from 13,000 to 10,800 BP.[13] A return of very cold conditions from 10,800 BP resulted in renewed valley glaciation in the Loch Lomond Stadial, which reached its maximum around 10,300 BP. In the Glen Roy area, ice advanced from the Great Glen eastwards up Glen Spean with a lobe extending 6 km up Glen Roy (to point E in the map above). Meltwater and runoff from unglaciated areas of Glen Roy/ Glen Spean became trapped in front of the valley glaciers and ice-dammed lakes formed. Stages in the retreat and downwasting of the glacier were marked by progressively lower ice-dammed lake levels in the Glens as successively lower cols became exposed and were able to act as lake outlets. The famous 'Parallel Roads' – horizontal narrow benches extending many kilometres round these glens – are the product of shoreline erosion at these col-controlled lake levels. Photograph 84b shows clearly the three Parallel Roads of Glen Roy at altitudes of 350 m, 325 m and 260 m. In Glen Roy, massive drift deposits up to 80 m deep were laid down just behind the maximum ice limit (see the map above); these deposits acted as a barrier to drainage down Glen Roy after the lake level fell below 180 m (the height of the feature) and a small lake (at F) survived into post-glacial times.

The terraces of lower Glen Roy commence below this drift dam and formed part of a series of terraces developed in Glen Spean as the ice-dammed lake (by now restricted to the western part of Glen Spean) dropped to successively lower levels with continued wastage of the ice blocking the western end of the valley. The first eight terraces graded to lake level altitudes in Glen Spean of 113 m (Terrace 1) down to 71.5 m (Terrace 8), with each terrace extending further west than the previous one. Of these, only Terrace 2 has extensive remnants in lower Glen Roy, where it forms a high Terrace some 30–40 m above the present channel. Its sand and gravel composition suggests that it was produced by a braiding river. Terraces 3–8 may well have formerly existed in Glen Roy, but were eroded by the River Roy as it cut Terrace 15/16, which is an extremely important feature in lower Glen Roy, with paired remnants more or less continuously from the drift dam to the Spean confluence (photograph 83 and the map). The fact that Terrace 15/16 grades close to the original altitude of the drift dam indicates that the River Roy did not incise the drift dam significantly until the whole of Glen Spean and Glen Roy had been deglaciated as Terraces 9–14, confined to the lower Spean, grade to Loch Lochy levels. The delay in downcutting may reflect the lack of coarse debris (the tools of incision) in the water issuing from the small drift-dammed lake above. Once the lake disappeared, due to infill from above by the upper Roy and gradual downcutting at its exit, the River Roy would have been able to incise quickly to form the current gorge section through the drift dam. Terraces 15/16 and the lower Terraces 17 and 18 represent successive early stages in the downcutting process. Together with a series of later lower terraces, these form flights of terraces which around Bohenie (photograph 83 and the map) are particularly well preserved.

The terraces of upper Glen Roy were not considered by Sissons, but photographs 84a and 84b demonstrate them to be major and extensively preserved features, being particularly prominent at the confluence with the River Turret, below the confluence with Allt na Rheinich and below the confluence with Allt Brunachain, with in each case the highest terrace over 30 m above the present river channel. A comprehensive terrace chronology has yet to be derived, but the persistence of the drift dam and its associated lake well into the immediate post-glacial, while Terraces 1–18 were produced in lower Glen Roy and Glen Spean, implies that the upper Glen Roy terraces developed in response to base levels controlled by the drift dam and lake in middle Glen Roy and are completely distinct from the lower Glen Roy terraces. The terraces presumably developed first in response to successive retreats of the ice-dammed Glen Roy lake as it fell from its 260 m Parallel Road position; then (and from the downvalley extent of the high terraces, most importantly) in response to the level of the drift-dammed lake remnant, which was initially 180 m;[14] and later in post-glacial times to successively lower levels of the River Roy as it incised the drift dam. The upper Glen Roy terraces would also be composed of glacial deposits relating to the main Devensian epoch (pre-13,000 BP), whereas the lower Glen Roy terraces are composed of deposits (till, fluviglacial and lacustrine) relating to the Loch Lomond Readvance.

This account indicates how complex explanatory accounts of river terraces can be and how terrace sequences along different reaches of the same river (the Roy) can be unconnected and have different explanations. In glaciated Britain, it is perhaps then to be expected that accounts of river terraces will be specific to individual river stretches and not easily correlated between river systems.

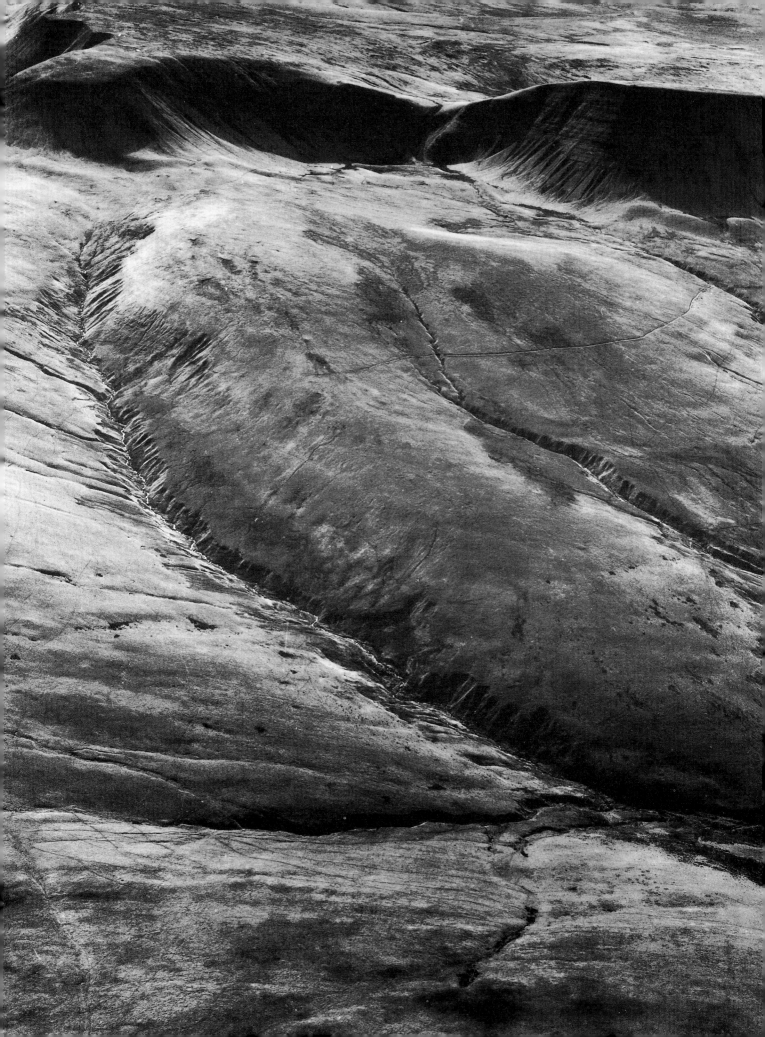

Aerial photographs in river terrace studies are mainly of use at the reconnaissance stage, as the required accuracy in height and slope determination in mapping terrace remnants usually necessitates ground survey, particularly as terrace height differences are normally much less than experienced in Glen Roy. Oblique photographs, however, if taken from the right direction with a low angle sun can highlight (via shadows) very subtle changes in slope and altitude indiscernible at ground level. This can, as in the case of the river Lugg near Kingsland (photograph 81), aid the identification and mapping of very minor terrace features.

Spatial variations in the distribution and depth of glacial and fluvioglacial deposits in glaciated areas have exerted (and continue to exert) a major influence on drainage density, on the nature and distribution of post-glacial fluvial erosion and on the character of rivers further downstream. The presence and available depth of unconsolidated drift (often added to by post-glacial peat development) represent a potential source of easily erodible sediment, which can be intensely gullied and deeply incised, whereas areas devoid of or covered by only a shallow veneer of drift are difficult to incise and gully and yield little sediment for transport by rivers.

The Brecon Beacons upland in South Wales has been mapped according to type and intensity of gullying[15] (see table and map below), with about 80 per cent of the surface area above 457 m displaying some form of gullying. On drift, gullying was probably most intense in immediate post-glacial times, when it would have been aided by meltwater and unvegetated surfaces; indeed many gullies were probably initiated as meltwater channels. However, gullying on the peatlands clearly post-dated peat formation and is considered to have commenced much later, around 2500 BP, with the change in climate from relatively dry sub-Boreal to a wetter sub-Atlantic type and an associated vegetation change from sphagnum bog to cotton grass moor. The headwaters of the Sawdde (photo-

85 *opposite* Post-glacial valley excavation in deep glacial till by the Nant Melyn and other headwater tributaries of the Sawdde in the Brecon Beacons; looking SSE. Deep incision has resulted in innumerable landslips close to the channel. These supply the bouldery bedload of the Sawdde and the unvegetated landslip scars act as important sources of suspended sediment, via rainsplash and frost-related processes. Also visible on the interfluves are shallow gullies (left) and numerous discontinuous pipes, the latter appearing as dark, discontinuous linear features (middle right). Nant Melyn at SN 807237, looking SSE.

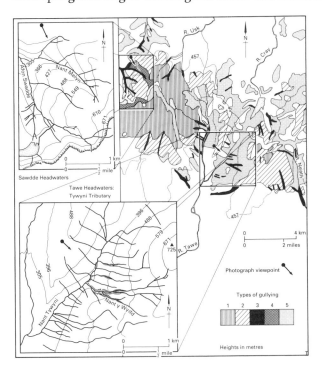

Distribution of gullying types in the Fforest Fawr section of the Brecon Beacons upland and gullies of the Sawdde and Tywyni headwaters.

Types of gullying in the Brecon Beacons

TYPE	DESCRIPTION
1	Peat-covered upper slopes with irregularly dendritic patterns of gullies which have reached a mature stage of development.
2	Peaty tracts, showing numerous shrinkage cracks or minor irregularities, but in which gullying is largely in the incipient or very youthful stage.
3	Deeply incised sections of main streams showing frequent slumping of high banks of bouldery drift.
4	Steep slopes of glaciated valleys with close systems of parallel shallow gullies.
5	Lower spurs or valley slopes with open systems of shallow gullies largely coincident with permanent minor lines of drainage.

86 Exceptionally deep post-glacial gullies incised in glacial till in the Nant Tywyni headwater tributary of the Tawe in South Wales. Drainage density, at 7.7 km/km², is exceptionally high by British standards. Note the old course of the railway on the valley-side, under which the gullies have been culverted. SN 865181, looking SE.

graph 85) and the Tywyni tributary in the headwaters of the Tawe (photograph 86) are classic examples of incision of very deep glacial drift (Gullying Type 3). Slumping is evident along all the main gullies. These slump features, which can themselves develop tributary gullies, play a twin role in current sediment supply to the Tawe and Sawdde. The slumps and landslips supply large calibre material to form the coarse bedload of the rivers. The unvegetated landslip scars subsequently act as major suspended sediment sources, via the processes of rainsplash erosion, ice-needle growth and other frost processes, and lateral corrasion by the river itself. Also evident in photograph 85 on the interfluves are peaty tracts with minor cracks and incipient gullies (Gullying Type 2); these are in fact areas of discontinuous piping, which is frequently encountered in peaty upland areas of Britain, and these tunnels can guide gully erosion at times of drainage network expansion. Great contrasts in drainage density occur in glaciated upland areas as a result of contrasts in degree of gullying. Drainage densities of individual 2×2 km^2 squares within the Brecon Beacons on Old Red Sandstone,[16] for example, range from 2.03 to 7.85 km/km^2. The intensely dissected Tywyni area (photograph 86) has a drainage density for OS square SN 8618 (see the map on p. 163) of 7.7 km/km^2. This certainly stems in part from the great depth of till available for dissection. Drainage network extension by gullying tends to be self-reinforcing as each gully incision creates additional highly erodible steep gully slopes enabling further gullies to be initiated by pipes, rills and seepages. However, there are many ungullied areas of deep till in the Brecon Beacons and gully initiation appears to require a trigger in many cases. The human factor remains largely unassessed, but it is feasible that many gullies may have been initiated during Medieval forest clearance and represent enlarged drainage ditches; modern parallels are gullies up to 6 m deep which have quickly developed from afforestation drainage ditches in the Towy headwaters in Wales.[17] In the Tywyni area, gully development has probably been exacerbated both by the construction of the railway, under which the gullies are culverted (photograph 86), and by historic mining activities in the eighteenth and nineteenth centuries.

CURRENT RIVER CHANNELS AND RECENT CHANNEL CHANGE

River channels in Britain exhibit considerable diversity in channel pattern, cross-sectional form, long profile and slope, and rates and nature of sediment transport and channel shifting. This variability reflects contrasts in catchment characteristics (notably relief, lithology and soils), catchment hydrology, sediment availability and type, and inherited valley character (materials, valley slope and valley width), which together interact to determine channel character. The Pleistocene legacy, which continues to exert a strong influence on many of these factors, varies considerably between catchments and is responsible for many of the contrasts we observe today in British rivers. Furthermore, many rivers, in adjusting themselves during the 10,000 years or so of the post-glacial period, have developed or retained features which, although stable in the present humid temperate climate, would not have evolved otherwise.[18] River character in Great Britain is thus a product of particular sequences of environmental conditions rather than a fixed set of current controls. The complex history of human impact on catchments and channels over the past 2,000 years in particular reinforces this need for a sequential or historical approach to understanding river channels.

This section concentrates on channel patterns and their recent history. British rivers have been classified[19] using sinuosity and activity as criteria:

1 active changing channels of low sinuosity, generally with braiding tendencies;
2 actively meandering rivers, that are perceptibly eroding the outside of bends and depositing point bars at the insides of bends; and
3 rivers with inactive, unchanging channel patterns, with few if any point or braid bars at low flow – this class includes stable meandering rivers as well as straight and irregular ones.

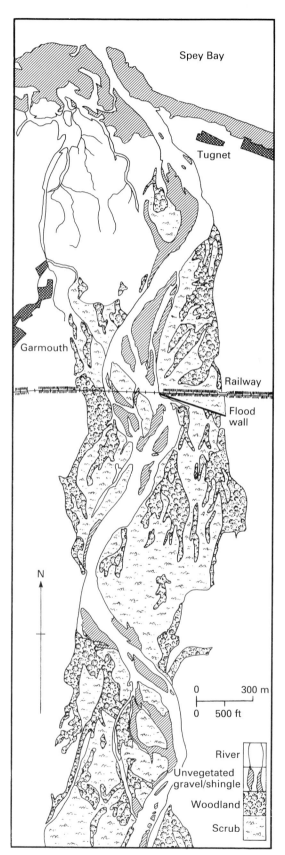

87 Channel and floodplain features of the braided lower Spey, Scotland. Note the classic features of a gravel-bed, braiding river: transverse and lobate central bars; vegetated islands (former gravel bars); lateral bars with chutes on the inside of meander bends. Note also the less frequently flooded scrub and woodland zone, the channels of which are activated during high flows of the river. The third zone, now cultivated or under planted forest, is only rarely flooded and has not formed part of the active channel for over 200 years; however, traces of numerous infilled and abandoned braided channels are clearly evident. NJ 343610 to NJ 341657.

Morphology of a braiding river; the lower Spey.

Although many upland rivers with gravel and boulder-sized bedloads exhibit braiding tendencies with prominent mid-channel bars causing division of flow at lower discharges, the only large British river characterised over much of its course by a high degree of braiding is the Spey, which drains much of Highland Scotland (photograph 87). The reasons for the extensive braiding lie in its high stream power, flashy regime with peak discharge up to 900 m³/s, a steep gradient (4.4 m in 1 km) right down to its mouth, and the availability of abundant coarse bedload material from Pleistocene till and outwash being reworked by the river and its tributaries. Lewin and Weir[20] used vertical aerial photographs, together with field survey and analysis of historical maps and documentary sources, to map the channel and floodplain of the lowest reaches of the Spey and assess their recent history. Stereopairs of 1:22,000 vertical photographs of 1967 were analysed using a Kern PG2 topographic plotter yielding floodplain maps with a spot height accuracy of ±0.4 m mean square error and a plan accuracy of ±1 m on the ground. This 1967 map was compared with historical surveys of 1760, 1876 and 1882–9 and a field survey of 1975. Three distinct zones, clearly identifiable on the photomosaic of photograph 87 and mapped in the accompanying figure, were distinguished:

(a) Present active channel. This is braided, but with a dominant meandering channel of low sinuosity at low flows, with emergent gravel bars (but no point bars) on the inside convex bends of the main channel. Sloughs or chutes tend to occur between the bank and the bar, thus distinguishing them from point bar features of meandering rivers. The degree of braiding is high at low flows, because of the emergence of the bars, and at high flows, because of the re-occupance of slightly higher channels around the vegetated islands, but least at intermediate flows.

(b) Zone of scrub and woodland worked over by the river in the last 200 years. Trees tend to occur on the finer sediments and areas abandoned by the river for the longest time. The vegetated islands are former gravel bars. This zone is fluvially inactive at lower flows, but is drowned and reactivated at high flows with net sedimentation in the channels.

(c) Remaining valley floor. This zone, though now cultivated or under planted forest, is still liable to inundation and sedimentation in higher flood events, but it has not formed part of the active channel for at least 200 years.

The degree of braiding in the lower Spey is less now than in the nineteenth century, with a braiding index of just 2.26 in 1967 compared with values of 5.00–7.40 in the surveys of 1760–1889. Successive aerial photographs from 1958 to 1975 show some changes in channel location, but no change in degree of braiding. The reason for the decline in braiding index may be due to one or more of a number of factors, ranging from human interference in the form of afforestation of part of Zone b and bank protection works, to a decline in flood frequency as a result either of land use change or a reduction in heavy rainfall magnitude-frequency.

In the Feshie tributary of the Spey successive vertical aerial photograph coverage from 1946 to 1967 was used by Werrity and Ferguson[21] to obtain a valuable insight into the temporal dynamics of braiding and the comparative roles of extreme floods and lesser, but more frequent flows in controlling braiding patterns. The Feshie, like the Spey downstream, has a very steep gradient (0.01) and is very flashy, with a mean flow from 1951–74 of 8.1 m³/s and a highest recorded flow of 200 m³/s. Braiding is well-developed in three reaches along the Feshie, but high terraces have constrained floodplain development and braiding along the remainder of its course. At the investigated braided reach by Glen Feshie Lodge, the river occupies a partially braiding and slightly sinuous channel shallowly cut into a terraced floor of coarse fluvioglacial outwash deposits (photographs 88a, b). As the photographs show, the terraces carry sparse remnants of Caledonian pine forest, but much of the floodplain is unvegetated.

The diagram (p. 170) shows changes in channel pattern as indicated by successive aerial photographs. In 1946, two distinct eastern and western channels dominated, with the link *BC* providing the main transfer of flow from the western to the eastern channel. Sinuosity of individual channels was quite high and braiding intensity was relatively low. By 1955 this pattern had become confirmed and simplified. The main western channel was more sinuous and less braided and between *B* and *D* had adopted a new sinuous course 70 m east of the 1946 one. The *BC* link to the eastern channel had been cut off by lateral bars, which had

(a)

88a, b The braided River Feshie at Glen Feshie Lodge before (a) and after (b) the flood of September 1961. Note the great increase in braiding which resulted from the 200 cubic metres per second flood, which reactivated some channels abandoned in the quiescent 1946–60 period as well as creating many new channels. NN 843934.

(b)

Changes in braiding pattern
and intensity of the River
Feshie at Glen Feshie Lodge at
successive aerial photograph
dates between 1946 and 1967.

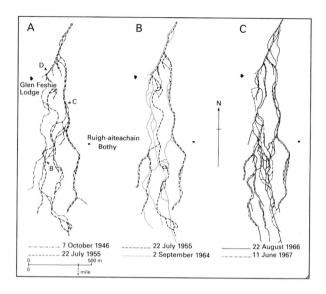

become stabilised by vegetation. Photograph 88a, taken in July 1960, shows this 'stable' pattern a year before the highest flow ever recorded – 200 m³/s on 28 September 1961 – obliterated the 'sinuous' 1950s pattern and replaced it with a highly fragmented and unstable braided pattern. Although the next series of verticals were not taken until 1964, an oblique of July 1962 (photograph 88b) shows the striking changes brought about by the 1961 flood and confirms that the single flood event was indeed responsible. Major effects of the flood included scouring of the floodplain, channel switching and re-excavation and use of former channels, and braiding on a scale more akin to a constrained proglacial sandur. The post-flood period 1962–81 witnessed a second period of rationalisation of the braiding pattern, accomplished by a combination of channel switching and channel migration via lateral bar development and meandering during lesser magnitude, but still competent flood events. Aerial photographs, used in combination with field survey work, thus demonstrate the dynamic nature of the current 'equilibrium' between an actively braiding river and its environment. The essential role of the occasional extreme event such as the 1961 flood on the Feshie in episodically re-establishing a very braided pattern is clear, as the lesser flood events tend progressively to reduce the degree of braiding.

Meandering by rivers in Britain is very common, but varies greatly in form, intensity and activity. Classic regularly-spaced meanders with smooth bends are comparatively rare and are usually found in wide alluvial plains composed of uniform alluvial material, where meanders are unconfined by valley or terrace walls and are uninterrupted by 'random' substrate variations. Photograph 89 shows some beautifully formed meanders along the Derwent in Derbyshire just prior to its confluence with the Trent. Sinuosity (the ratio of actual channel length to straight-line valley length) at 2.2 is very high. However, apart from a little bank erosion at the incipient cut-off (SK 432326), there has been no change in the meandering pattern at least since the first edition of the Ordnance Survey in 1836. This lack of activity, which is common to many meandering rivers in Lowland Britain, is also indicated by the lack of any prominent visible point bars in the photograph. Whereas previously such patterns have been viewed

as stable forms in equilibrium with the current environment, more recent thinking considers the meandering to have been inherited from previous more powerful rivers, with present day flows and gradients insufficient to accomplish significant bedload transport or bank erosion and hence incompetent to change the inherited pattern.

In contrast, many other meandering rivers are characterised by actively changing courses and fluvial geomorphologists have used aerial photographs and topographic maps of different dates to investigate modes and rates of meander migration and changes in sinuosity through time. The River Dane between Congleton and Holmes Chapel in Cheshire is an example of an actively meandering

89 Inactive meanders of the River Derwent, Derbyshire, close to its confluence with the River Trent. Despite a very high sinuosity (2.2), the meanders show very little change since 1836 apart from some erosion of the neck of the incipient meander cut-off. SK 431327.

90a, b Active meandering by the River Dane between Congleton and Holmes Chapel, Cheshire: (a) section CB near Swettenham; (b) section ED west of Radnor Bridge; in both cases the river is flowing WNW from right to left. Traces of former courses including those of 1839–41 (see the diagram on p. 173) are evident in both stretches as the river has migrated and become increasingly sinuous over the past 140 years. Sinuosity in reach CB has increased from 1.82 to 2.67 and in reach ED from 2.06 to 2.31 between 1839–41 and 1984. (a) SJ 800666 to SJ 791672; (b) SJ 830651 to SJ 819651.

(a)

(b)

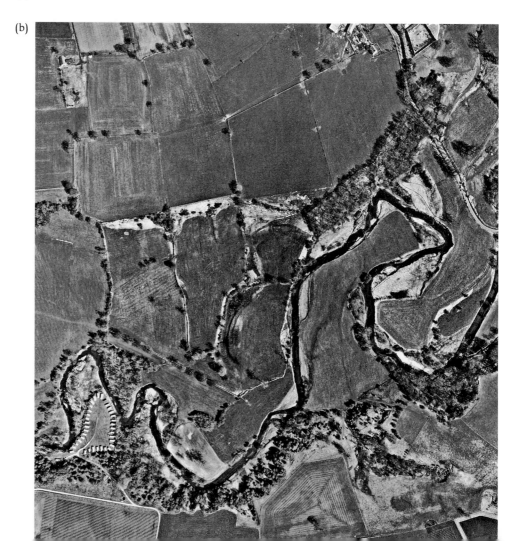

Channel change along the meandering River Dane, Cheshire, between Holmes Chapel and Congleton from 1842 to 1975.

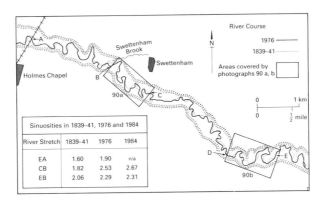

Sinuosities in 1839–41, 1976 and 1984			
River Stretch	1839–41	1976	1984
EA	1.60	1.90	n/a
CB	1.82	2.53	2.67
EB	2.06	2.29	2.31

river with tortuous, irregular meanders occupying the full width of (and sometimes confined by) the rather narrow floodplain. Comparison of Ordnance Survey maps of 1842 (based on ground surveys of 1839–41) and 1976 (1:25,000 Second Series) indicates some major meander migration and development, with an increase in overall sinuosity of the section between Radnor Bridge and Holmes Chapel from 1.60 in 1839–41 to 1.90 in 1976 (see the diagram above). An examination of aerial photographs of two sections showing greatest changes confirms

91 Current and past meanders of the River Glass, Scotland, between Glassburn and Struy. The river is flowing NNE from bottom to the top of the photograph. Comparison with old maps shows that the present, gently meandering course (sinuosity = 1.25) has not changed since the late nineteenth century. However, the photograph shows clear evidence of a much more sinuous channel in the past. At least eight cut-off meanders are clearly visible, with ox-bow lakes preserved in two of them. The meanders have much tighter bends and lower wavelengths than the present meanders, but the channel width appears relatively unchanged. NH 375352 to NH 399393.

the validity of these changes, with the old courses still clearly visible and active meandering in the form of unvegetated point bars and undercut banks readily discernible (photographs 90a, 90b). Increased meandering is particularly evident in the Swettenham reach between points *B* and *C* (photograph 90a) where sinuosity increased from 1.82 in 1839–41 to 2.53 in 1976 and 2.67 in 1984 (the date of the photograph). The increase in sinuosity in section *DE* west of Radnor Bridge (photograph 90b) is more modest (from 2.06 to 2.31 between 1839–41 and 1984), but some parts of the channel have migrated over 100 m from their former positions.

However, aerial photograph evidence of past channel courses can be misleadingly 'fresh' and does not necessarily imply that the current river is actively shifting its course. A stretch of the River Glass (photograph 91) between Glassburn and Struy shows very clear evidence of a much more sinuous channel in the past. In the 5 km stretch covered by the aerial photograph, there are at least eight abandoned meander loops, most of considerably lower wavelength and much more tortuous than the present meanders. Two of the abandoned loops contain ox-bow lakes – Loch an Deala (NH 394382) and Loch an Eilein (NH 381351) – and a channel width not dissimilar from that at present is also indicated by the abandoned channels. However, comparison of the 1975 aerial photographs with the first Ordnance Survey map of 1873 indicates that the current channel is, except for a few minor lateral bars, identical in pattern to that of 1873 and thus can hardly be classed as 'actively meandering'. Indeed, the absence of any marked point bars or lateral bars in photograph 91 tends to confirm the 'inactive' nature of the present River Glass. That the more sinuous channel predates 1873 is clear, but when and why the sharp reduction in sinuosity from around 2 to the present 1.25 occurred is unknown. One possibility is that deforestation either in medieval times or at the time of the 'Agricultural Clearances' (dispossession of crofters in the Scottish highlands, dating from 1750, by landowners who introduced sheep-rearing) may have resulted in a sudden increase in flood magnitude-frequency and that the associated increase in stream power resulted in a series of meander cut-offs and a much less sinuous channel. A modern parallel is the River Bollin in Cheshire where a series of meander cut-offs and an associated decline in sinuosity from 2.34 in 1935 to 1.40 in 1974 was attributed to an increase in flood magnitude-frequency resulting from urbanisation and land drainage schemes in the catchment.[22]

THE HUMAN IMPACT ON RIVER SYSTEMS

Human activities have had a profound and increasing impact on drainage basins and river channels. Forest clearance and wetland reclamation for agriculture, subsequent land use changes, mining activity, industrialisation, urbanisation, land drainage schemes and direct interference with rivers through reservoir construction, channel straightening, bank stabilisation and floodplain building have all influenced river channel form and process on a range of timescales. Because of the length of time over which human activities have extended, their impact is very difficult to isolate from natural fluvial adjustments to post-glacial conditions and climatic changes. Thus, with reference to the previous section, whether the

current braiding activity of the Feshie and Spey is natural or to some extent a consequence of historical land use change, whether the meander cut-offs and decline in sinuosity of the River Glass were due to an extreme hydrological event or a sudden land use change, and whether the recent active meandering by the Dane is normal or has been enhanced by the increased flows and sediment supply of an agricultural catchment are all unclear. It is often very difficult to interpret present process and forms in a longer-term context.

The consequences of recent human activities have been better documented and have formed a major theme of fluvial research during the past two decades. Aerial photography has proved an invaluable aid in many of these studies, both in helping establish the 'pre-impact' river channel state (particularly as studies are

92 Burrator Reservoir and the wooded River Meavy, Devon; looking north. A reduction in flood-magnitude frequency since reservoir construction has led to channel deposition and a reduction in channel capacity in the Meavy downstream. Looking N from SX 554673.

The impact of reservoirs on downstream channel capacity; the contrasting effects of Burrator Reservoir on the Meavy and Camps Reservoir on Camps Water.

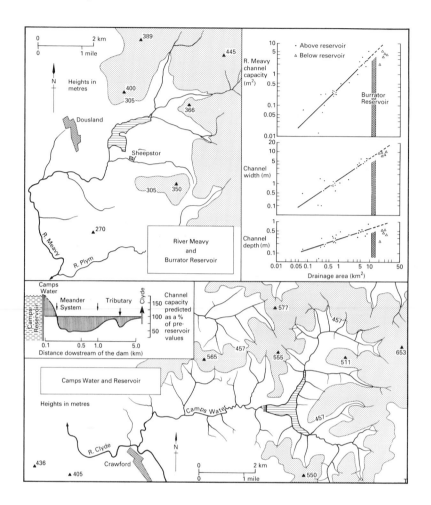

often instigated *after* the problem has been caused!) and in monitoring or dating the sequence of river channel changes following human actions. This section examines five examples of different types of recent human impact on British rivers.

Reservoirs significantly alter the flow and sediment regimes of rivers downstream of them reducing the size and frequency of floods and cutting off the supply of bedload and suspended sediment from upstream. However, the effects on the river channel downstream vary from river to river both in scale and nature. Burrator Reservoir in Dartmoor (photograph 92) has resulted in a sharp reduction in channel capacity (cross-sectional area) downstream along the River Meavy, mostly achieved through a decline in channel depth rather than width.[23] Channel capacities at the five monitored cross-sections downstream of the dam were 27–62 per cent of the pre-reservoir values predicted by extrapolating the channel capacity/basin area relationship derived from upstream cross-sectional data. Camps Reservoir on the Camps Water tributary of the Clyde in the Scottish borderlands has produced a rather more complex channel response.[24] The channel immediately below the spillweir flume at the dam has experienced channel enlargement rather than reduction and channel capacities at individual cross-sections have increased by over 80 per cent. Post-reservoir discharges in this case are clearly still competent and the absence of a load from upstream has resulted in scour of the downstream channel. However, within 250 metres of

the dam, the river enters a series of active meanders (photograph 93), within which the channel capacity (mostly through a reduction in width) has been reduced to around 46 per cent of expected values (see the diagram on p. 176). A renewed reduction in capacity occurs below the first significant tributary, as the reduced peak flows of the main stream cannot move all the sediment supplied by the tributary and it is deposited immediately downstream of the confluence. The absence of initial channel enlargement by the Meavy downstream of Burrator probably results from its wooded, steep, bedrock course being less erodible than the alluvial sediments downstream of Camps Reservoir, as a comparison of photographs 92 and 93 demonstrates.

The River Ystwyth in west Wales has been affected by human activities in several ways. The river, which has been investigated in detail by J. Lewin over many years, is characterised by a rapidly shifting, gravel and boulder bed channel, with numerous lateral and mid-channel gravel bars (photograph 94). The high bedload is mainly the result of post-glacial reworking of bouldery till and outwash by the river and its tributaries, but has to some extent been enhanced by medieval forest clearance and extensive mining activity in the catchment. Human activity has, however, significantly interfered with the course of the lower Ystwyth on a number of occasions in the past 125 years with major repercussions on channel behaviour and the pattern of floodplain development.[25] Railway construction in the 1860s involved the building of an embankment and artificial straightening of a 4 km stretch of river above the bridge at Llanilar. Historic maps

93 Camps Water downstream of Camps Reservoir, Scotland, looking east to the dam. The reservoir has resulted in an increase in channel size over the first 250 m downstream of the dam, but a decrease in capacity in the meandering stretch further downstream. Post-dam peak river flows, though reduced in magnitude-frequency, are still competent as the initial channel capacity increase resulted from scour and the meanders have continued to migrate. NT 001225.

show the re-establishment of a sinuous pattern during the succeeding century, but the railway embankment confined the meanders to the northern part of the floodplain. Photograph 94 clearly shows the large expanses of formerly active channels south of the railway embankment, now protected not only from channel encroachment but also from flooding. The sinuous channel was re-straightened in 1969, but without bank protection works, and this enabled the stretch to be used as a 'natural laboratory' by Lewin, who monitored fluvial re-adjustments in detail until late 1970 when the River Authority once again re-straightened the channel. In 1969–70, as a consequence of the increased slope and stream power which artificially straightening a meandering river inevitably involved, erosion and sediment transport in the reach was greatly increased and the following sequence of development was observed. First, a series of regularly-spaced, transverse mid-channel bars developed; these resulted in bank erosion at the same regularly-spaced intervals. The mid-channel bars tended to become attached alternately to either bank and to develop into point bar features with pronounced erosion on the opposite bank. Once the sinuous channel had been created, however, flow patterns and bank erosion became less dependent on initial forms, with further lateral accretion becoming a result rather than a precursor of bank erosion. A three-phase development was thus recognised:

1 'Initiation': transverse bar formation and resultant bank erosion.

2 'Development': cut-bank development, resultant point-bar sedimentation and unit riffle-pool development.

3 'Dismemberment': series of alternative phases such as (a) multilooping, (b) neck of chute cut-offs and (c) impingement of non-adjacent channel reaches.

94 *opposite* The River Ystwyth and its floodplain above Llanilar Bridge in 1975. The river is flowing from east to west. The old railway embankment, running parallel to and just south of the river, was constructed in the 1860s and has prevented the river (which was straightened in 1864, 1969 and 1970) from migrating freely over its floodplain. Pre-1864 channels south of the embankment are still clearly visible and heavy metal concentrations are highest in these floodplain sediments, corresponding with the peak in mining activity upstream at that time. Note the central and lateral gravel bars and the meandering main channel which have developed in the present river since it was last straightened in 1970. SN 619757.

In the Ystwyth example, following a total of only 75 hours of flows competent to move the gravel-bed material in the single winter of 1969–70, thalweg sinuosity had risen to nearly 1.2 and bank erosion totalling 50 per cent of the original channel area had occurred. Photograph 94 shows the stretch of river in 1975 after it had been re-straightened in 1970. Many of the features noted by Lewin had re-developed, with large longitudinal gravel bars, a meandering thalweg, some point-bar development in the upper stretch and a series of small lobate and other mid-channel transverse bars in the lower stretch. Channel straightening without channel bank protection clearly results in rapid channel sedimentation, widening and shifting in a river with such a high bedload and frequent high flows as the Ystwyth.

River pollution has increasingly become a focus of attention of fluvial geomorphologists. The distribution of heavy metal pollution in the floodplain of the lower Ystwyth has been shown to reflect the interplay of floodplain and mining history.[26] Mining for lead and zinc within the Ystwyth catchment extends back to the Middle Ages, but became intensive in the nineteenth century before declining as prices fell in the 1890s. Mining processes were relatively inefficient at that time and wastes contained a high proportion of ore particles and metal-rich spoil. Mining inputs to the Ystwyth included sediment and solutes from crushing mills, dressing floors, flotation plant, spoil tips and mine drainage waters from mining complexes at Cwmystwyth, Pontrhydygroes, Frongoch and Grogwynion.

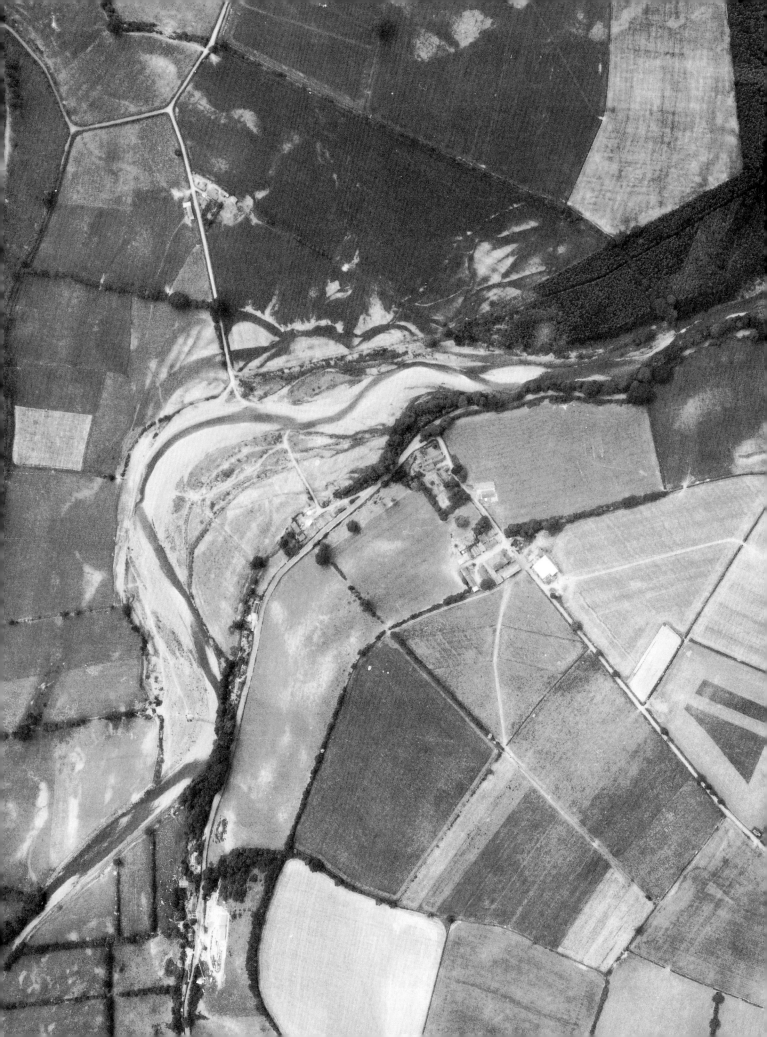

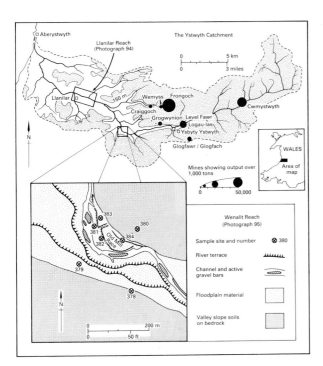

Sources of mining pollution and monitored areas of polluted floodplain sediments in the Ystwyth catchment and location of sampling sites at the Wenallt site.

The pattern of floodplain pollution is closely related to the position of the river channel at different times in mining history. Analyses of floodplain sediment samples at different points along the Ystwyth, including Llanilar (photograph 94) and Wenallt (photograph 95) demonstrated this strikingly, as shown in the table on p. 273. At Llanilar, highest metal concentrations were found in old channel sediments south of the railway embankment on the site of the active channel at the height of mining activity immediately prior to railway construction in the 1860s. At Wenallt, where a meander loop has shifted within a band of about 100 m over the last 150 years, highest values were recorded in old sediments of the nineteenth century away from the present channel on the inside of the meander bend and in the bed of a mill race, abandoned by 1904, where more toxic fine material had tended to accumulate. At both locations, sediments of the active channel and contemporary floodplain are characterised by significantly lower, but still serious metal levels, reflecting on the one hand a sharp reduction in pollution inputs because of the collapse of the mining industry during the first part of this century, but at the same time demonstrating the long-term nature of metal pollution in a situation like the Ystwyth where the river is continually shifting its course and re-working polluted floodplain material.

Although the scope for using aerial photography in studies of river pollution (and natural river water quality) is limited by the process nature of the topic, there are occasions where aerial photographs can help identify and quantify pollution sources and controls. Since the early Industrial Revolution, the River Tawe has been a major source of heavy metal pollution, a consequence of the concentration of metal smelting works in the lower Swansea Valley. Pollution levels in the lower Tawe and its right bank tributary, the Nant-y-Fendrod,[27] have declined considerably since the 1960s (when systematic data were first collected by the River Authority), but remain very high, as shown in the table on p. 273. The

95 *opposite* A migrating meander loop of the gravel-bed River Ystwyth at Wenallt. The river is flowing from east to west (right to left). The history of mining in the catchment is reflected in variations in heavy metal concentrations in the floodplain sediments. Metal concentrations are highest on the partly vegetated older sediments north of the meander bend and in the Old Mill Race, parallel with the road, both last active at the end of the nineteenth century. SN 675715.

Changes in the industrial
landscape of the Lower
Swansea Valley between 1965
and 1986 for the area shown in
photograph 96a.

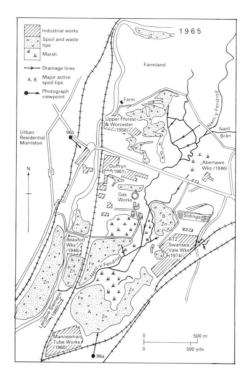

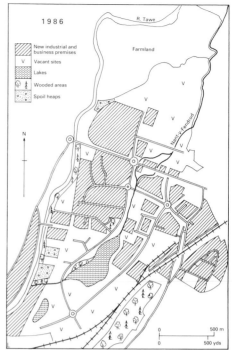

(a)

(b)

main reasons for the decline in pollution levels are (1) the collapse of the metal smelting industry, with the last works closing in 1974, (2) extensive concrete culverting of the Nant-y-Fendrod system so as to avoid old works and contact with spoil material and (3) the large-scale re-development and reclamation from dereliction of the Lower Swansea Valley. There has been a complete transformation of the landscape of the Lower Swansea Valley (see diagram on p. 182), involving flattening of spoil heaps, the development of a new industrial area, establishment of proper roads and drainage, landscaping with trees and grassland, re-alignment of the Nant-y-Fendrod and construction of a flood relief channel and lake. With the disappearance of the old landscape, the use of pre-redevelopment maps and, even better, aerial photographs become essential in attempting to examine and explain past pollution levels and trends.

Photographs 96a and 96b, taken in 1962, indicate clearly some of the principal sources of heavy metal pollution at that time. Two massive spoil tips to the north and southwest of the Rio Tinto Zinc Works dominated the 1962 landscape, with numerous smaller tips and waste surfaces also evident. Both the large spoil heaps showed evidence of gullying and were clearly being undercut by the Nant-y-Fendrod stream. Overland flow from the spoil areas and works complexes and gullying and lateral stream erosion of the large tips would clearly have been significant contributors of particulate metal pollution to the stream during storm peaks. This, together with scour of metal-rich sediments along the bed of the stream, probably explains why particulate heavy metal pollution levels were highest in winter storm peak events.[28] Because some of the particulate load readily passes into the dissolved state, dissolved metal concentrations in the streamwater also tended to exhibit peaks in high winter flows. Highest dissolved concentrations, however, occurred (a) at very low flows in summer, when effluent from the metal works (when still operating) and metal-rich subsurface runoff from the tips were least diluted by the low river flows, and (b) in the rising limb of summer storm hydrographs that followed long dry spells, when 'flushing' occurred of accumulated, readily soluble material on spoil and works surfaces and in drainage ditches by overland flow and streamflow.

It is interesting and significant that whereas mean levels of zinc, cadmium and lead have fallen sharply since the late 1960s (see the table on p. 273), maximum levels have continued to be high into the mid-1980s. This perhaps reflects the fact that the land reclamation programme itself (at its peak in 1981–6) involved bulldozing and large-scale disruption of spoil material and hence, temporarily at least, increased exposure of fresh metal-rich material to physical entrainment and solution by overland flow and percolating water in storm events. The decline in mean concentrations, on the other hand, reflects baseflow conditions, where the increased culverting of streamflow so as to avoid contact with spoil and the lack of metal-rich effluent from the now defunct metal works have resulted in major improvements in baseflow quality. With re-development now nearly complete (see the diagram on p. 182) and the disruptive phase over, questions remain about future pollution levels as mean concentrations are still unacceptably high. Little is known about likely long-term trends in subsurface water quality of seepage through the now redistributed metal spoil material which covers much of the Swansea Valley floor. Also, the flood relief lake may be acting as a short-term

96a, b *opposite* The River Tawe, the Nant-y-Fendrod tributary and sources of heavy metal pollution in the Lower Swansea Valley in 1962. (a) looking north, with a massive zinc/iron tip complex in the foreground (SS 673964) and another zinc-rich spoil tip in the top right (SS 675977), both undercut by the Nant-y-Fendrod. Surface erosion by overland flow and rainsplash, subsurface solution by percolating water and direct fluvial erosion by the Nant-y-Fendrod, together with direct industrial effluent, were the main mechanisms by which heavy metals entered the river system. Note the new straightened, culverted channel as well as the old, meandering channel of the Nant-y-Fendrod; both were used at this stage. (b) Looking southeast from Morriston, with the Duffryn Metal Works in the left foreground and a large zinc-rich spoil tip, actively impinged by the old course of the Nant-y-Fendrod, with the Rio Tinto Zinc Swansea Vale Works behind it.

97 *opposite* The upper Towy valley in pre-reservoir and pre-afforestation times in July 1960. The whole of the east (right) side of the valley is now under mature conifer forest. Note the afforestation ditches on the southern interfluve of the LI3, Nant-y-Craflwyn catchment (at SN 812508). The deeply incised tributary further north is LI2, the Nant-y-Fannog, where the mature conifer forest has recently been subjected to bankside clearance and liming in the Llyn Brianne Acid Waters Project. SN 809515.

sink for metal pollutants and become a pollution hazard in the long-term. With the development of a yacht marina at the mouth of the Tawe and a planned Tawe barrage, these are important questions. Using short-term field data, heavy metal and other pollution events in the Tawe river system have been modelled with some success using a time series approach[29], but the long-term applicability of such a model in such a rapidly changing landscape as the Lower Swansea Valley is questionable, given the parallel changes in pollution sources and contributing processes indicated above.

The hydrogeomorphological consequences of different types of upland land use have formed a major theme of fluvial research over the past two decades.[30] Recently attention has focussed on the impact of upland land use and land management practice on streamwater acidification. The Towy headwaters around Llyn Brianne Reservoir (photographs 97–100) have been the subject of some detailed investigations in the 1980s. Afforestation of considerable sections of the area commenced at the end of the 1950s (photograph 97) and has been extended since (see the map below and photograph 100 on p. 188). The dam and reservoir (photograph 99) were completed in 1972. Recent work by the Welsh Water Authority[31] showed that acidity and aluminium levels in many of the streams in the upper Towy catchment are episodically very high and that the streams could not support fish and had depleted population of aquatic plants and animals. Furthermore, problems appeared more acute in afforested than in grassland catchments. Also, the problem of acid streamwaters appeared to be widespread in the extensive area of upland Wales underlain (as at Llyn Brianne) by inert Ordovician and Silurian rocks and characterised by acid, often peaty soils and streamwaters of

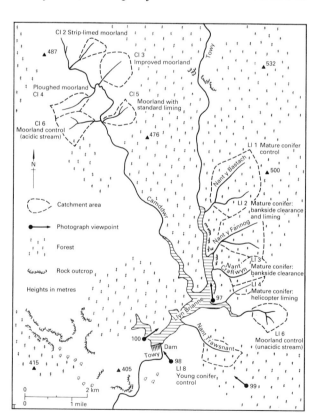

The Llyn Brianne Acid Waters Project; monitored catchments in the Upper Towy catchment, Central Wales.

very low background hardness (and hence low neutralising capacity). A DOE/ Welsh Office-funded research programme commenced in summer 1984 to investigate further the effects of different types of land use on streamwater acidity. Thirteen small catchments were selected for long-term study, 5 acting as 'controls' of different land use type and 8 being used to assess the impact of a range of land management treatments. The project aimed therefore at identifying practical ways of alleviating the acidity problem as well as seeking scientific explanations.

A pilot survey in September/October 1984 of three control catchments of contrasting land use demonstrated the much higher and serious acidity and aluminium levels of the conifer catchments, set out in the table on p. 273, and that conditions were worst during flood peaks, when surface and near-surface runoff processes make their greatest contribution to streamflow. Oblique aerial photographs show the positions of the three catchments, in the cases of the conifer catchments prior to afforestation (photographs 97, 99, 100). These results have been confirmed by continuous records of water quality and flow from mid-1985 onwards. Worst episodes of high acidity and aluminium tend to occur in high flow events following long dry spells, particularly in autumn when dead vegetal matter and associated acids are flushed out of catchments, and following snowmelt in winter. For example, on 4 March 1986 a 50 mm rainfall combined with melting

98 *opposite* Llyn Brianne Dam on the River Towy. Note the concrete spillway. The dam was completed in 1972. SN 792485, looking NW.

99 *above* Llyn Brianne reservoir and dam. The northwest flowing stream in the right foreground is the Nant Trawsnant (LI8) in 1972 prior to afforestation. In the Llyn Brianne project, it is the Young Conifer control catchment and is characterised by high acidity and aluminium levels, though lower than those found in mature conifer catchments. The pre-afforestation drainage net can to some extent be checked from the photograph. SN 792485 looking NW.

of snow lying since the end of January resulted in a pH fall in catchment LI1 (mature conifer) from 6.0 to 4.0 and in LI6 (unacidic grassland) from 7.1 to 5.4 in the succeeding runoff peak.

However, it is dangerous to draw simple conclusions concerning land use – and afforestation, in particular – and streamwater quality from the catchment water quality results alone. Soil type and slope runoff processes vary greatly within the Llyn Brianne area with topography and superficial deposits (peat, glacial till, soliflucted till) and differences in streamwater quality between catchments partly reflect the varying proportions of different topography/soil/hydrology units within each catchment. Field surveys of variations in water quality within individual catchments demonstrate this clearly. The high pH of moorland catchment LI6 (photograph 100), with values normally between 6.0 and 7.5 and a minimum recorded pH of 5.2, obscures the fact there are some very acid pipeflows and tributaries within the catchment. The northern part of the catchment contains several pipes in acid peat with pH values typically within the range 3.95–4.98 and the northern tributary has pH values of less than 6.0 even at baseflow in winter. The more dominant tributaries in the south of the catchment, however, drain slopes covered with deep glacial till, which because of its unconsolidated nature is more susceptible to weathering and hence buffering of the slope waters which drain through it. The high pH values of the southern tributaries (6.7 at winter baseflow and over 7.0 in summer) and a large pipe in slumped glacial till (5.6 to 6.3) contrast with their more acid northern counterparts.

There are several mechanisms which may account for higher acidity and aluminium levels under conifer forest than under grassland. These include: (1) more effective entrapment of air pollutants by the taller and aerodynamically rougher forest canopies than by lower, more streamlined grass or moorland vegetation; (2) acidification effects of some conifer species on stemflow and water dripping from the forest canopy; (3) acidifying effects of deep, acid, conifer needle litter on infiltrating water; (4) increased uptake by trees of limited calcium and magnesium available in the soil, thereby depleting the soil's buffering capacity and leading to increased leaching of aluminium instead; (5) effects of forest soil profile development on soil permeability and soil chemical reactions; (6) effects of deep ploughing and drainage ditch construction prior to afforestation on slope hydrology and runoff hydrochemistry by inducing greater speed and amounts of shallow soil drainage (which bypasses deeper, more base-rich, buffering horizons of the soil) and disruption of iron- and aluminium-rich horizons typical of many moorland soils; (7) effects of increased evapotranspiration and reduced runoff under forest compared with grassland in producing less dilution of acid-rich drainage. The relative importance of each of these possible factors is yet far from clear, though results to date suggest (1) and (6) to be of great significance.

Aerial photography is proving useful to the project in a number of ways. Because different topography/soil/vegetation units are characterised by contrasting slope hydrology and runoff hydrochemistry, precise maps of these variables for each catchment are essential in explaining catchment water chemistry. Large-scale vertical air photography is therefore being used in soil and vegetation mapping and digitised terrain analysis. It is also being used to delimit the drainage

100 *opposite* Southern part of Llyn Brianne reservoir. The unacidic moorland LI6, Nant Esgair South catchment lies in the top right of the photograph. SN 822496, looking ENE.

network, including rills and drainage ditches, and to identify and map areas prone to overland flow and soil pipe systems of various types. The existence of oblique photographs of some of the catchments in pre-afforestation days is particularly useful, as it allows the pre-afforestation drainage network and catchment condition to be assessed and compared with the current situation at least on a qualitative basis. Thus photograph 97 shows the young conifer catchment LI8 prior to afforestation and photograph 98 was taken while afforestation was actually in progress in what is now a mature conifer catchment (LI3) with the pattern of deep ploughing in the south of the catchment clearly visible. The utility of aerial photographs as historical sources is once again evident.

GUIDE TO FURTHER READING

K. J. Gregory (ed.), *River Channel Changes*, Chichester, 1977.

G. E. Hollis (ed.), *Man's impact on the Hydrological Cycle in the United Kingdom*, Norwich, 1979.

D. K. C. Jones, *The Geomorphology of the British Isles: Southeast and Southern England*, London, 1981.

D. Knighton, *Fluvial Forms and Processes*, London, 1984.

J. Lewin (ed.), *British Rivers*, London, 1981.

K. S. Richards, *Rivers: Form and Process in Alluvial Channels*, London, 1982.

6 Coastal landforms

The coastline of Britain displays a great variety of geological and morphological features.[1] The coast can be defined as a zone extending seawards of the present limit of wave activity (the cliffline in many areas) to the water's edge at low tide, while the landward limit may be marked by the top of a high cliff, the head of a tidal estuary or the inland margin of a series of sand dunes. The coastal zone can therefore vary considerably in width, and it is a dynamic, mobile environment where both erosion and deposition occur as a result of the interaction of many natural processes.[2]

Cliffs are eroded by the direct impact of waves, the corrosion and abrasion of loosened material, frost action and the chemical weathering of the rocks. Biological processes may also contribute to the weakening of rock strata, especially where limestone bedrock is involved. Landslides and rockfalls also contribute to the recession of cliff faces, sometimes on a very large scale and involving many acres of land and thousands of tons of weakened rock strata. The rate of loss of land by erosion varies enormously, from perhaps only a few millimetres per year on the 'hard', resistant rocks of north-west Scotland, to several metres per year on the relatively 'soft' rocks of south-east England. Substantial accretion of gravel (the term shingle is commonly used when the material is well-rounded), sand and mud has also occurred around our coasts, giving rise to beaches, sand dune systems and salt marshes. In areas protected by wave-constructed shingle beaches and spits and in estuaries, quiet water conditions can be found where muddy sediments can accumulate. Salt-tolerant plants often aid this process and give rise to extensive salt marshes to the limit of high spring tides.[3]

Wind is probably the most important of all the climatic factors directly involved in coastal evolution in Britain. The strong winds generated by the frequent passage of low pressure systems across the country directly affect the sizes and patterns of the waves reaching the coastline. Wave action, together with currents and the rise and fall of the tides, determine the circulation patterns of near-shore waters, influencing the type of deposition of coastal sediments and the manner in which erosion takes place. Although Britain is situated in an exposed position, where the considerable fetch of the North Atlantic Ocean allows large powerful wave systems to develop with the aid of the wind, there is considerable variation in their effect because of the protection afforded by the enclosed nature of the North Sea and the Irish Sea to some parts of the coastline. There is also variation in the tidal range of normal tides which can markedly influence the height range of beaches and salt marshes, even though Britain as a whole is thought to enjoy an environment with a universally large tidal range; strong coastal currents capable of moving sand-sized sediments are generated particularly in places where the tidal range exceeds a few metres.

Location of aerial photographs
in chapter 6.

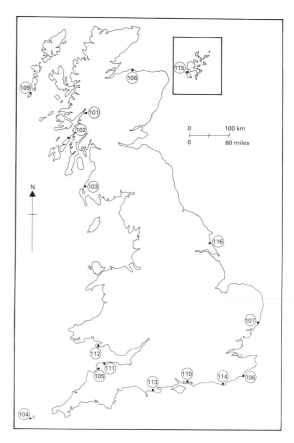

Extensive sand dune systems have been constructed in localities where a moderate to large tidal range exposes a wide sandy beach at low tide which supplies the sand to be blown landwards. Particular types of vegetation can then 'fix' the sand although there is still considerable mobility among many of Britain's coastal dunes. Some of the larger dune areas appear to have resulted, at least in part, from particularly windy, and possible stormy conditions during the fourteenth and fifteenth centuries, when serious encroachment by blown sand occurred, for example, in Cornwall, South Wales and Scotland (see chapter 7).

In addition to the important rise and fall of the tides, which acts within a well-defined height range, it is the more spectacular changes of level between land and sea throughout geological time that have been so important in influencing the evolution of our coastal scenery. Traces of former Tertiary marine activity is believed to extend to +200 m, while so-called 'interglacial shorelines' have been recorded up to +30 m, and some of the associated marine deposits may be up to several hundred thousand years old.

However, it is the sea level changes since the maximum of the last glaciation (Devensian), some 18,000–20,000 years ago, that are particularly important.[4] Over much of the northern half of Britain the weight of the ice masses depressed the land, which recovered to approximately its former level following complete ice wastage about 10,000 years ago. This comparatively local 'up-and-down' movement of the land is termed the isostatic factor. The growth and expansion of the world's great ice-sheets (in Antarctica, Greenland, North America and Scandinavia), and the more limited growth of ice masses in the British Isles, brought about a lowering of world sea level: these world-wide 'up-and-down' movements in sea level are referred to as the eustatic factor. In the southern half of Britain it is the eustatic control of sea level which appears to have been so important, whereas in the northern half of the country very complex changes of both land and sea level took place as a result of eustatic and isostatic factors. The picture is further complicated because of the tectonically-controlled down-warping of south-east England, the southern North Sea, and adjacent continental areas.[5] However, it is clear that as deglaciation proceeded there was a steady encroachment by the rising Flandrian sea in southern Britain across the exposed continental shelf, to carry vast quantities of sediments, which have eventually been added to our beaches and sand dunes. In marked contrast, in northern Britain *both* the isostatic and eustatic factors operated. Complex changes took place, the sea at one time advancing across areas still depressed from the weight of the original ice masses, at another being forced to withdraw as land recovery continued at a more rapid rate than the eustatic rise of sea level. These so-called marine transgressions and regressions have been responsible for sequences of elevated or raised shorelines and a variety of marine deposits found both above and below the present shoreline.

The following photographs and text will examine, first, coastal features where significant sea level changes have been involved; second, features resulting from the accumulation of shingle, sand and muddy sediments; and third, a variety of cliff forms.

SEA LEVEL CHANGES AND COASTAL LANDFORMS

An example of a partially drowned coastal landscape is provided by Loch Leven in the Western Highlands of Scotland (photograph 101). Loch Leven is a tributary valley of Loch Linnhe, the latter being the south-western end of the Great Glen 'trench' which extends across mainland Scotland to Loch Ness and the Moray Firth in the north-east. These are sea lochs, and represent land-locked fjords with floors down to 200 m below sea level. This part of the Scottish Highlands has been severely glaciated to produce a rocky, ice-scoured landscape and highly indented coastline (cf. photographs 5 and 32). Loch Leven extends back eastwards between mountain summits exceeding 600 m in height and the photograph shows summits in the distance rising to over 900 m. Rock types consist mainly of Pre-Cambrian quartzites and mica schists, all strongly folded and faulted and with major granite and volcanic intrusions. From these mountains ice discharged westwards and south-westwards to overdeepen valleys such as Glencoe and to gouge the fault-guided depressions in the ancient and hard rock formations to considerable depths below present sea level. This occurred at a time when the sea had withdrawn because of eustatic lowering consequent upon the build-up of the world's great ice-masses. The floors of these glaciated trough valleys or fjords are irregular, consisting of a series of rock basins, separated by rock sills, upon which varying amounts of sedimentary fill have been left by the ice, and subsequently added to by the rivers, and finally drowned by the transgressive sea.

The last ice mass to form a significant major ice cap over these mountains developed thicknesses ranging from 400–600 m, and virtually the whole of the

landscape depicted in the photograph must have been covered by ice. This ice advance is known as the Loch Lomond Readvance (or Stadial) which took place only some 11,000–10,000 years ago and represents a re-occupation of a small part of the area covered by ice during the maximum of the last glaciation (Devensian) some 18,000–20,000 years BP. However, at the time of the Loch Lomond Readvance the forward margin of the ice reached near Ballachulish, just to the foreground of the photograph. Enormous quantities of outwash sands and gravels were deposited at the ice margin and in such volume that the mouth of Loch Leven is still restricted by the sediments. The tongue of land extending across the Loch from the north side consists mostly of these old glacial outwash deposits.

The final major landscape-forming agency has been the sea, for in this area there has been considerable isostatic adjustment of the crust as the ice has decayed, together with the eustatic recovery of sea level.[6] Complex adjustments between land recovery and sea-level rise permitted two marine transgressions of parts of south-west Scotland. The first is known as the Main Late-Glacial Shoreline, when the rising sea was able to flood the still depressed landmass, at the time of the Loch Lomond Readvance. Beyond the position of the ice front in Lochs Linnhe, Leven and Etive this shoreline is now recorded at +11 m near Oban, but declines in height towards the west and south-west. These shorelines are developed as rock platforms at a number of sites, but are not recorded inside the outermost limit of the Loch Lomond Readvance ice caps and glaciers, thus indicating that they date from this time and were not formed earlier than 11,000–10,000 years BP.

The second shoreline represents the much later post-glacial transgression, and it too is elevated above present HWM, occurring at +12 to +14 m at the head of Loch Linnhe, and at the mouth of Loch Etive, and Loch Leven. Subsequently, residual isostatic recovery raised this shoreline. On the ground we see evidence that the outwash deposits at Ballahulish, like those in Loch Etive and Loch Linnhe

101 *opposite* Loch Leven, Argyll/Invernesshire, Western Scotland. Loch Leven extends eastwards from Loch Linnhe and is enclosed by ice-scoured mountains, the summits of which range from 600 m to over 900 m in height. This sea loch is a good example of a coastal fjord, with a submerged rock floor some 45 m below sea level off Ballachulish, and at Corran Ferry in the foreground. The entrance to Loch Leven has been narrowed by the deposition of fluvioglacial sediments of Loch Lomond Readvance (Stadial) age. Raised shorelines of late- and post-glacial age form narrow coastal terraces at +10 m to +14 m. NN 053598.

Loch Leven, West Scotland
A. The drowned glaciated valleys or sea lochs ('fjords') of Linnhe, Etive and Leven.
B. The extent of glaciation during the Loch Lomond Readvance (Stadial).
C. The isobases of the Main Post-glacial Shoreline illustrating the differential isostatic recovery in Scotland during the last 6–7,000 years.

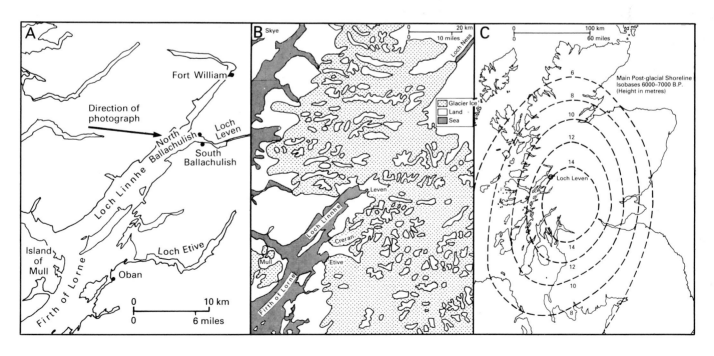

102 Jura, Inner Hebrides, Western Scotland. In the foreground there is an elevated shoreline with the relic or 'dead' cliff cut in resistant Pre-Cambrian (Dalradian) quartzite rock. The notch at the base of the 'dead' cliffline is at +7 m to +10 m. The lateral extent of the elevated rock platform and its considerable width (up to 150 m) should be noted. The entrance to a former sea cave can be seen in the lower centre of the photograph and relic sea stacks project above the general level of the rock platform in the right foreground. NR 655970.

have been partially trimmed and cliffed during this transgression. Thus, in places the narrow coastal terrace extending around the coast and *into* Loch Leven is regarded as being an elevated marine feature dated to about 6,000–7,000 years BP. Ballachulish has in part been built upon this marine terrace, but it is the deeply incised mountain-girt fjord which dominates the coastal landscape.

In the Inner Hebrides an elevated shoreline in the form of a rock platform achieves a maximum altitude at the cliff notch of +7 m to +10 m OD (photograph 102). Detailed examination suggests that in Jura, Islay and Mull this platform has a gradient of 0.13 m/km to 0.16 m/km to the south-west.[7] The elevated marine-cut feature is shown here developed in hard Pre-Cambrian quartzite on the north-west coast of Jura. The rock platform may be 20 to 30 m wide, but can reach 150 m, and the cliff at its rear rises sharply from a well-defined notch, and may be over 15 m high.

The origin of such features has attracted a number of hypotheses. That such a rock platform can be cut in hard, resistant rock is quite possible provided,

for example, that sufficient time and appropriate local conditions of exposure to storm waves are satisfied. It is not surprising that such an extensive feature should therefore have been regarded as either preglacial or interglacial in age and thus considerably older than any of the late-glacial and post-glacial elevated shorelines recorded in Scotland. A post-glacial age is generally rejected because there is no evidence of platform development in rock at this level where such planation can be proved to be younger than suites of glacial deposits exposed along the coast.

However, the evidence that this rock platform has been glaciated, which would confirm its preglacial or interglacial age, is now regarded as doubtful. The presence of erratics or patchy glacial till resting on its surface has been accounted for by slumping, or by floating pack-ice lodging deposits on the rock, and the widespread presence of true glacial striae is now denied. Furthermore, the existence of delicate stacks and arches on the platform, in areas of intense glacial erosion during the last glaciation, indicates the strong possibility of a more recent origin. This view is enhanced because the platform is also present in relatively sheltered areas, protected from strong wave action and where the fetch is minimal. Nevertheless, enormous volumes of rock must have been removed to account for the existing width of these platforms, often backed by high cliffs. The photograph indicates that the elevated platform, stacks and cliff are remarkably fresh in appearance, and one of the most important clues to their age is that nowhere does the platform occur inside the limits of the Loch Lomond Stadial ice limit (see the map on p. 195). There is a possibility therefore that the platform was cut by the sea at the time of this ice advance, when marine processes were aided by powerful weathering of the rock by the freezing and thawing of sea water. The efficiency of a combination of marine abrasion and rock shattering by freeze–thaw cycles at the cliff base has been amply demonstrated on present day polar coasts, especially where cliff recession has been aided by frost-riving of the cliff face, the resulting debris being removed even by limited wave activity.[8]

The Western Highlands and Inner Hebrides were of course still considerably depressed isostatically between 11,000 and 10,000 years ago, probably by at least 40 m below today's level. With sea level recovering rapidly at this time from its low level during the glacial maximum, the sea would have been able to mount a limited transgression against the part of the coast not covered by the ice of the Loch Lomond Stadial. Elsewhere it would have been in contact with glaciers flowing into some of the western sea lochs (e.g. in Loch Long, Loch Fyne, Loch Linnhe). Although pack-ice and ice-foot may have protected some segments of coast, damping down wave action, elsewhere, even the hard quartzites of Jura would have been unable to resist the combined attack of waves and frost action. Later, as the Loch Lomond Stadial ice wasted away isostatic recovery brought about the elevation of the platform and the relic or 'dead' cliffline.

Submerged peat beds and occasional tree stumps in position of growth are present at various levels below high water mark at many localities in Britain, and seen at times of low water and in harbour excavations. Their dating by pollen analysis and radiocarbon assay indicates that a marine transgression took place during the Holocene Epoch, which began about 10,000 years BP. The extent and timing of the marine transgression depended upon the relationship between the rate of rise of sea level and the isostatic movement of the land area resulting from the decay of the ice masses. Sometimes local barriers of glacial deposits, together with marine-constructed sand and shingle bars, modified the progress of the transgression. Consequently, the main Holocene marine transgression is not a synchronous event around the coastline.

Near Ballantrae, and north of Benane Head on the Ayrshire coast there is an elevated shoreline with 'dead' cliffs (i.e. cliffs beyond the reach of modern wave action), a shore platform and shingle bars (+7 to +10 m OD), marking the main post-glacial raised shoreline (photographs 103a and b, and the accompanying map). The cliffline is for the most part cut in glacial till, even though at many coastal sites in south-west Scotland the location of the main Holocene shoreline is closely related to the position of an older rock platform and associated cliffline. Thus in places exhumation of an older marine-cut notch has taken place, but here the relic cliffs are steep (slopes of 25°–38°) and they are almost entirely in glacial till. The cliffs have receded from what was probably a nearly vertical profile, mainly as a result of weathering and severe gullying of the glacial deposits, which is particularly well illustrated in photograph 103b. At Ballantrae the River Stinchar reaches the sea through a complex of raised shingle bars, and the outlet has changed position during the last few hundred years as a result of the interplay of fluvial and marine processes.

103a, b *opposite* Ballantrae, Ayrshire, South-West Scotland. An elevated shoreline with 'dead' or relic cliffs is shown in the photographs, where a well developed shore platform and notch is cut in glacial drift. This is an example of the main post-glacial raised shoreline; in places the drift has been removed to reveal an older marine-cut notch and cliff in rock, of unknown age. Photograph (a) is taken looking north-east across Ballantrae village and the mouth of the River Stinchar. Photograph (b) is taken looking south-west towards Balsalloch Hill and Benane Head. a. NX 083825; b. NX 093865.

(a)

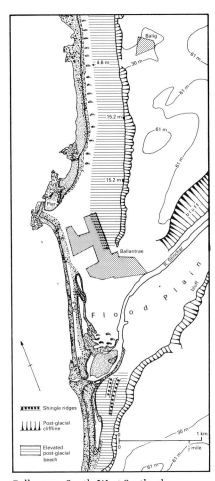

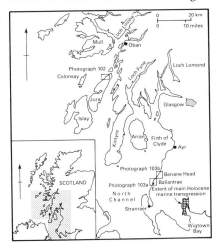

Ballantrae, South-West Scotland
The elevated shoreline of the main post-glacial marine transgression is shown near Ballantrae. The flood plain of the River Stinchar is enclosed by steep bluffs, which have resulted from the migration of the river and undercutting of the glacial sediments and bedrock: these bluffs pass almost imperceptibly into the relic sea cliffs to the north and south of the village.

(b)

104 *opposite* Isles of Scilly. The photograph is taken looking from north-east to south-west across White Island, the western part of St Mary's, to St Helen's, Tern, Tresco and Bryher. The archipelago is seen to consist of a series of islands, islets and reefs of varying size and shape, and all are composed of granite bedrock, overlain in places by some glacial and periglacial sediments, together with blown sand. SW 930155.

Isles of Scilly
A. The relationship of the granite masses of Land's End and the Isles of Scilly.
B. Major rock structural trends, distribution of granite tors and land over 30 m.
C. The former ice limits indicated by outwash gravels and distribution of periglacial head; blown sand and a raised shore platform (often with beach gravels).

The Holocene marine transgression penetrated several kilometres inland at the head of Wigtown Bay and Luce Bay, and near Troon and in the Glasgow area, but was restricted to less than 100 m along much of the coastline. Near Ballantrae the transgression began shortly after 8,400 years BP, although it started as much as 1,000 years earlier along the Solway Firth. In both areas the transgression had culminated by about 6,500 years BP, although at the head of Wigtown Bay it may have continued until about 5,000 years BP. The emergence of the landmass and the consequent regression of the sea from the main Holocene transgression level of approximately +10 m OD is marked by lower shoreline features at +5 to +6 m OD, and may be represented by shingle bars in the foreground of photograph 103a. At the head of Wigtown Bay such features have been dated at about 2,000 years BP.[9] In sheltered areas of the Solway Firth, the Firth of Clyde and the Firth of Forth (and Forth Valley), extensive marine–estuarine sedimentation took place during the Holocene transgression. The sediments laid down at that time now form widespread raised mudflats or 'carselands'.

Evidence for eustatic changes of sea level can be seen in the Isles of Scilly, which consist of some 140 islands and islets of mainly granite bedrock. These islands provide an excellent example of a drowned archipelago where the highest points seldom exceed 60 m. They rise from a larger and presumably once continuous granite outcrop intruded into the older Palaeozoic (Devonian?) basement rocks, and this outcrop is related to the other granite masses of Land's End, Bodmin Moor and Dartmoor. The landscape and coastline as we see it today represents the culmination of the drowning of a dissected landmass, where wave action has exploited joint planes and other fracture systems in the bedrock (photograph 104).

The effects of the Flandrian sea level rise to about its present position, some 4,000 years ago in West Cornwall, would have permitted the submergence of

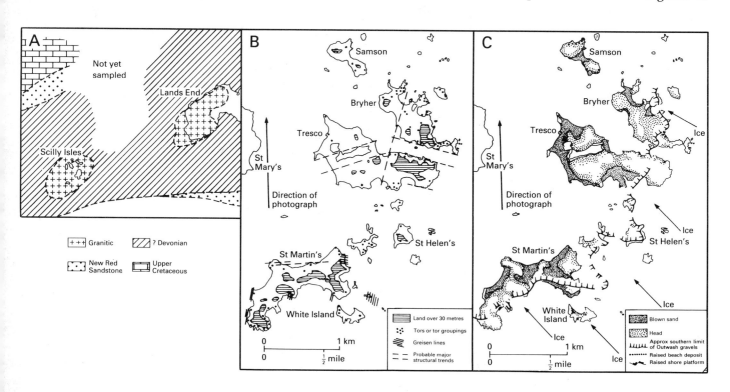

(a)

105a, b The Taw–Torridge Estuary, North Devon. The oblique photograph (a) is taken looking westwards across the joint estuary of the Rivers Taw (on the right) and Torridge (on the left), from above Instow, in the foreground. The sedimentary deposits in the estuary are displayed at low water, while the sand dune systems of Northam and Braunton Burrows can be seen to the south and north of the estuary respectively. The vertical photograph (b) *opposite* provides detail of the mouth of the estuary, with The Bar marked by the arc of white surf and with numerous ebb and flood channels. SS 463307.

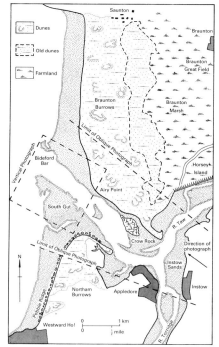

Taw–Torridge Estuary, North Devon
The Taw–Torridge estuary forms an important break in the cliffline of north Devon, between Saunton Down to the north and the high ground to the south of Appledore and Westward Ho! Extensive deposits of marine and aeolian sediments can be seen to confine the present exit of the rivers Taw and Torridge to Barnstaple Bay.

202

(b)

any low-lying granite areas (reduced by sub-aerial erosion since exposure of the granite) and the creation of the archipelago. With world sea level still rising at a rate slightly in excess of 2 mm/year in West Cornwall, the Isles of Scilly will clearly retain their characteristic features. The coastline is varied, ranging from extensive sandy strands to low castellated granite cliffs, with tors, deep weathering features and clifflines developed in glacial drift and periglacial head deposits which form substantial coastal terraces.

An important ice-limit has been mapped as lying across St Martins to Tean, St Helen's, Tresco and Brynher and it is represented by till deposits and extensive outwash gravel deposits. The age of this ice limit is still the subject of controversy, with Mitchell and Orme[10] suggesting a 'Wolstonian' (penultimate) glaciation age, whereas more recently Scourse prefers a Devensian (last glaciation) date.[11] Disintegration of the granite has provided considerable volumes of sand to nourish the beaches and dune systems (see map above).

In sharp contrast to the archipelago of the Isles of Scilly the joint estuary of the Rivers Taw and Torridge (photograph 105a) forms an important break in the high cliff line of north-west Devon. The dual river outlet is restricted by the growth of the large sand dune systems of Northam Burrows and Braunton Burrows (see map below). These sand dunes have accumulated as a result of the construction of shingle bars, the most notable of which is at Westward Ho! Within the shelter of these barriers, large areas of mudflats and salt marshes have contributed to the restriction of the river channels, and some have been reclaimed (e.g. Braunton Marsh).

The mouth of the drowned estuary, where enormous sedimentary deposits now obscure the bedrock profiles of the original river valleys, is shown in photograph 105a. In fact, the rock channel just north of Appledore lies at −25 m and at the mouth of the joint estuary −30 m and extends seawards an unknown distance. Such buried rock channels are known throughout the West Country and south Wales and indicate one or more of the stages of glacial low sea level. Recovery of the sea from the low levels attained during the last glaciation brought about the transgression of former land surfaces, indicated by submerged forests at many sites including Westward Ho!, Saunton and within the estuary itself. The rate of sea level recovery slowed down dramatically after about 5000 years BP, when it was still some 5 m *below* present sea level. During the next 3,000 years the sea gradually encroached upon the landmass, which remained unaffected by the isostatic rebound recorded in the northern half of Britain.[12] The transgressing Flandrian sea progressively submerged the coastal margin, penetrating far inland to form this large estuary, and in so doing re-worked a host of superficial deposits, contributing large volumes of sand and gravel against the new coastline to nourish the present beach and dune systems. Both photographs were taken at low water when a series of large sand and gravel banks can be seen, with prominent ebb and flood channels, which constantly change position. The vertical aerial photograph 105b displays the mouth of the main channel in more detail and the outer arc of breaking surf marks a distinct area of shallows, known as The Bar. Such zones of shallowing are well known at river mouths where copious quantities of sediments are available.

CONSTRUCTIONAL, COAST FORMS: SHINGLE, SAND AND MUD

Constructional features form major contrasting morphological features of the English coastline. For example, Dungeness and Orford Ness are large shingle structures resulting from the accumulation of considerable volumes of shingle and sand, constant wave action, and involving some minor changes of sea level over the last 5,000–6,000 years or so. Coastal dune systems also form conspicuous and important features of many segments of the coastline and all stages in their growth and decay can be observed. From low mounds of sand accumulating on the upper part of a beach they can grow to form foredune ridges and, provided first, that sand supply is adequate, second that erosion is restricted on the seaward side, and third that vegetation grows sufficiently rapidly, then large dunes can form. The importance of marram grass (*Ammophila arenaria*) in helping to stabilise new sand is well known, and with time and variation in sand supply, changes in the vegetation cover develop, the dune colour gradually alters from white to grey, reflecting increased stability and gradual development of soil. Two examples have been chosen to illustrate the development of dune systems – the Culbin Sands on the Moray Firth, and a beach–dune–machair complex on the island of Barra in the Outer Hebrides.

Where shelter is provided by shingle and sand barriers, then fine mud and silt suspended in the water can be deposited. Layer after layer of fine sediments can, with the aid of salt-tolerant plants, build into extensive mudflats and salt-marshes, thus bringing about considerable progradation of the shoreline. The example chosen shows the marshes and mudflats of Langstone Harbour on the Hampshire coast.

Dungeness foreland is unique, for it constitutes the greatest individual spread of shingle in Britain, extending from the old abandoned coastal port of Winchelsea to Hythe. It is a complex structure made up of a very large number of individual shingle ridges (photograph 106), in part over a base composed of fine sands and grey muds. Each ridge appears to represent a stage of growth of the foreland as it extended south-eastwards into the English Channel. Over 90 per cent of the shingle is flint, which must have been derived from the erosion of the chalk cliffs west of Eastbourne, and from the floor of the English Channel. There is a small amount of material derived from the Upper Greensand (cherty sandstones) and from the Hastings Beds (sandstones), presumably from the eroded cliffs near the present site of Hastings. There are a few far-travelled pebbles of quartzites and coarse grits, which owe their origin to movement up the English Channel from the west and to the dumping of ballast from ships in the vicinity of the present coastline.

The processes operating today are chiefly those initiated by wave action produced by the prevailing south-westerly winds. The 'southern face' of the ness lies almost perpendicular to approaching wave fronts from that direction, while the 'eastern face' has been formed by waves produced by winds from an easterly quarter. Thus the plan shape of the ness is maintained, with shingle being moved from the southern side, round the point, where it is constructed into low shingle ridges. Some shingle may be forced back southwards temporarily by wind-driven waves from the east, but in the main the ness appears to be stable and to be

Dungeness, Kent
A. Stages in the evolution of the Foreland.
B. Detail of the distribution and pattern of the shingle ridges of which Dungeness is largely constructed.

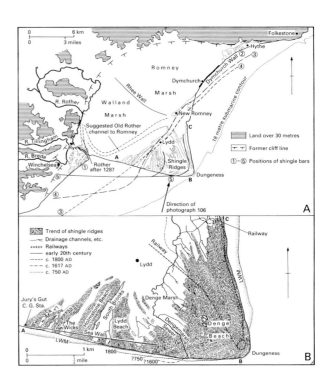

106 Dungeness, Kent. The most distinctive features of Dungeness Foreland, photographed from south-west to north-east, are the complex patterns of the shingle ridges and the general symmetry of the foreland as it projects into the English Channel. Romney Marsh extends beyond the shingle ridges in the distance, while there is an indication of the presence of deeper water (over 18 m) in the foreground and at the tip of the foreland as compared with the shallower waters to the east of Dymchurch. TR 093168.

moving slowly north-eastwards as there is loss from the exposed shingle bars between points A and B and accretion between B and C, as shown on the map opposite.

There is a steep, ridged profile leading from the level where the power stations and lighthouses are built to high water mark, and the position of the 10 fathom (submarine) contour (about −18 m) indicates that the ness has grown outwards into deep water. The steep beach profile seen near point B must therefore continue below low water mark. The movement of the whole ness towards the east is reflected in the re-positioning of the various lighthouses, beginning in 1792 and culminating in the most recent in 1960, each having to be placed a little further east as the point of the ness changed position.

While most of the processes operating today are understood and the change of position of the point of the ness over time can be measured precisely, the same cannot be said for the evolution of this massive cuspate foreland. The various theories of origin, notably that of W. V. Lewis, are reviewed comprehensively by J. A. Steers[13] and can only be outlined here with the aid of the map on the opposite page.

The events leading to the evolution of Dungeness begin with the existence of a shallow bay extending from Winchelsea to Hythe, and the old sea cliff has been identified at these sites and at Rye and Appledore. There followed a growth of shingle bars and spits, first at Camber Castle (1) (built in 1538–9) and near Hythe (2), and in addition a more or less continuous shingle ridge was developed by wave activity extending from south-west to north-east across the bay, from near Hastings to Hythe and Folkestone (3). Subsequently, other ridges were constructed, occasionally breached by waves and the River Rother, and slowly the tendency of the dominant waves from the south-west to re-orientate the shingle bars would have swung shingle bar 3 to position 4 and eventually position 5, when the shape of the ness began to take on the plan form we recognise today.

Groups of shingle bars known as the Midrips, The Wicks, Holmstone Beach and Lydd Beach, and then the ridges extending to the point of the ness, are believed to trace the progressive changes. The diversion of the course of the River Rother and its tributaries from an exit through the outer shingle ridge (at 3) near Romney to the present outlet near Rye may have been accomplished by breaching of the main shingle ridge by storm waves. Whether or not such breaching was assisted by small changes of sea level is debatable, but it is quite clear that the marshes have grown up inside the protecting shingle ridges and reclamation has been aided by the construction of the Rhee and Dymchurch Walls. The rate of accumulation of the marsh sediments is not yet known but there has been piece-meal embanking of Walland Marsh, south of the Rhee Wall (Roman?), mainly during the fourteenth to the seventeenth century but with some patches dating from the eighth to the thirteenth century.

The prominent shingle structure known as Orford Ness projects into the southern North Sea as a great foreland, and a shingle spit extends some 14 km to the south-west to bring about a major deflection of the River Alde near Aldeburgh and the River Ore (photograph 107). The former cliffline, with a base at 0 m to +8 m, which predates the growth of the shingle structures, can be traced from Aldeburgh to Orford, Hollesley and beyond. It is cut in early Pleisto-

107 Orford Ness, Suffolk. The prominent and massive form of the projecting 'ness' or headland can be seen in the distance, together with the long sweeping 'tail' of the spit with recurring shingle ridges in the foreground. The shallow estuary of the River Ore is shown to be completely submerged at high water, with the sea lapping at the outer edges of the protected coastal marshes of this generally low coast. In the foreground can be seen the shingle accumulation at Shingle Street, but the Bar, seawards of the entrance to the estuary, is submerged, and there is no visible indication of the mobile nature of the spit over the last two centuries. TM 422500.

cene Crag deposits (chiefly sands and gravels) overlying Tertiary London Clay as bedrock (see maps 1, 2 and 3 on p. 209). Extensive deposits of estuarine clays accumulated as the rising Flandrian sea encroached on the land area. The progress of this transgression is documented by the included peat beds, which indicate that marine sedimentation occurred from $8,460 \pm 145$ years BP (sample at -12.7 m to $3,460 \pm 100$ years BP (sample at -3.4 m) approximately. Thus, while the precise date for the beginning of the accumulation of the shingle foreland and spit is not known, it cannot have begun until about 3,000 years ago, when the sea achieved its present level.[14]

The changes in the configuration of the coastline over the last 300 years are illustrated from a selection of available maps and it is clear that considerable fluctuations have taken place both in the position of the distal end of the spit and in its width.[15] Recently, detailed investigations of the structure of the spit and movement of shingle have been undertaken, and it has been shown that the growth of the spit has indeed been as erratic as these maps suggest. Between 1945 and 1962 the average prolongation of the spit (at North Weir Point) was 20 m/yr, but growth was extremely variable and the build-up of the spit was not uniform.

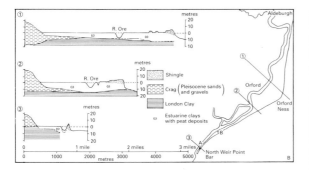

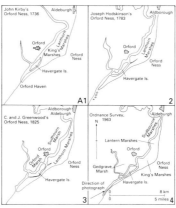

Orford Ness, Suffolk
A. 1, 2, 3, 4. The changing form of the Ness and spit is displayed in these maps of 1736, 1783, 1825 and 1963.
B. Cross-sections of the River Ore and Orford Ness showing the approximate extent of Tertiary (London Clay) and Quaternary (Pleistocene Crag sands and gravels and Holocene estuarine clays and peats) deposits.

While the direction of maximum wave fetch is from the north-east wave attack can occur from between north-north-east and south-south-west at North Weir Point. It has been shown that as growth has taken place, beach ridges immediately north of the tip of the spit have increased in height, largely by refashioning, replacement and the addition and removal of shingle. When the spit has been receding there has been a corresponding wave attack and loss of land at Shingle Street, with the River Ore maintaining its channel width. When, on the other hand, the spit was extending southwards then protection was afforded to Shingle Street, where shingle bars developed. The distal end of the spit is prevented from growing landward by the depth of the river channel (8 m below OD) and the tidal race at the mouth of the estuary.

Differential erosion and accretion has occurred in the southern part of the spit, leading to the straightening of the spit as it has extended. The marked concavity near point A began about 1921 during a period of retreat of the spit, but as a result of later accretion with the influence of waves from a southerly quarter, the spit straightened and increased in width, especially north of point A. Thus relatively rapid rates of movement of beach material have been shown to take place, and while *net* movement of shingle is north to south, much re-sorting has taken place many times as progradation and recession of the spit has occurred. New beach ridges have been added during accretion, merging with earlier ones north and south of the marked concavity (at point A), and thus apparently continuous ridges of mainly flint shingle may not be synchronous along their entire length, segments having been built at different times. Within the shelter of the spit, salt-marshes have developed, some of which have been reclaimed north of Orford, which was a port in the twelfth century. To the north, the seaward face of the ness is undergoing retreat between the point and Aldeburgh. It is not yet possible to say just what effect the continued slow subsidence of the crust in the southern North Sea will have, nor can the effect of the equally slow rise in world sea level be evaluated for this coastal area. If these trends continue at rates of a few millimetres a year, however, then further important coastal changes can be expected to take place.

Where no ready source of sediments such as glacial deposits or from rivers is available, and where the rocks are extremely resistant, another source of sand-sized material is provided in the form of millions of tons of shell fragments, as for example on Barra, in the Outer Hebrides. All coastal dunes in Britain result from the wind blowing across the inter-tidal areas which are exposed at low water, the rapidly dried sand being removed and carried inland. Some sand may form miniature dunes on the highest part of the foreshore, in amongst the shingle, and any seaweed or other flotsam carried to and beyond the spring tide level by wave action. Here certain annual plants such as sea rocket (*Cakile maritima*), prickly saltwort (*Salsola kali*), sea sandwort (*Honkenya peploides*), hastate orache (*Atriplex hastata*) and sea wormwood (*Artemisia maritima*) trap sand and, provided the winter storms do not succeed in removing the embryo dunes, further growth may occur. However, where the supply of sand is sufficiently great to permit accumulation then certain grasses are most important in promoting dune growth. Of these sand couch grass (*Agropyrum junceum*) is often a pioneer, to be followed rapidly by marram grass (*Ammophila arenaria*) the most important of the plants responsible for dune construction.

In Scotland where the two examples are located, the vegetation types chiefly involved in promoting coastal dune growth are marram grass, sea lyme grass (*Elymus arenaria*) and sand couch grass. Other plant associations take over the role of dune fixing as sand movement diminishes and the marram dies back. Strong winds blowing from a variety of directions permits dune shapes and erosional 'blowouts' to occur which deviate from the prevailing (westerly) wind direction, while excess precipitation can inhibit sand movement for many months. Very considerable diversity can be observed in coastal dune systems, ranging from old, fixed and completely vegetated (grey) dunes to young active foredunes (yellow dunes), where nourishment from the beach is still taking place, and where erosion may also be active in causing 'blowouts'. There are also areas of relatively flat or gently undulating sand surfaces where typical dune vegetation has been replaced by grasses, the so-called machair of the western coastlands.

A very large beach–dune system is found along the southern shores of the Moray Firth. The north-flowing Highland Rivers of the Spey and Findhorn have carried enormous quantities of sediments to the coast to supplement the glacial and fluvioglacial deposits left by the ice-sheets which flowed generally from west to east through the Moray Firth. Wave action has been responsible for the movement and construction of these readily available sediments into an extensive set of raised beaches (with the aid of isostatic adjustment of the landmass as the ice sheets decayed and finally disappeared) and a series of magnificent shingle bars and spits. Many of these shingle ridges have crests that rise a few metres above present HWM and they constitute a contemporary phase in the evolution of this section of coastline. It is upon these multiple shingle ridges that many of the sand dune systems have developed, the most famous of which is the Culbin Sands, where blown sand has encroached some 5 km inland.[16] The segment of coastline depicted in the two photographs (108a, 108b) and in the accompanying map extends from near Nairn eastwards to Findhorn and Burghead Bay. Afforestation of part of the extensive sand dune area had been completed when the photographs were taken, but in the seventeenth century rich agricultural land was to be found here. Huge dunes have been built by sand blowing from west to east leading to gradual accumulation, although violent storms immediately prior to 1695 may also have contributed to the inundation, and the process continued until afforesta-

108a, b *opposite* Culbin Sands, Nairn, Scotland. The photographs show a stretch of coastline between Nairn and Burghead Bay, where the extensive sand dune system of the Culbin Sands has formed, and which is now largely afforested. Extensive changes in the coastline have occurred over time, especially at the mouth of the River Findhorn (A), and at The Bar (B). Active dune systems testify to the continued mobility of the sand, where blowouts alternate with partially vegetated areas. NJ 042643.

210

(a)

(b)

Culbin Sands, Moray Firth,
Scotland
The Culbin Forest with coastal
features.

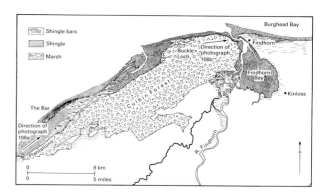

tion began in the 1920s. Although now largely covered by coniferous plantations, the Culbin sands were among the most extensive such dunes known in Britain. Some dunes are still active and carry a typical vegetation, mainly marram, and there has been limited encroachment of the sand onto areas with growing trees in several areas, notably around Maviston.

Considerable coastal changes have also occured at the mouth of the River Findhorn, which formerly continued its course westward via Buchie Loch. The postion of the present mouth of the river is attributed to the breaching of the deflecting shingle bar during a storm on 11 October 1702, which also accounted for the loss of part of Findhorn town. The river mouth is still being pushed westwards by spit growth near Findhorn.

Another prominent feature is The Bar, which is recorded by the 1835 Admiralty Survey (photograph 108b). It exhibits a mixture of sand and shingle accumulations that form a basement for some dune growth. At the south-western end, however, there is a series of prominent recurved shingle laterals displaying the direction of growth from north-east to south-west. It seems very likely that the whole structure is being displaced westwards along the coast by the dominant waves, and a rate of movement of about 1.6 km/century has been suggested. Landward of The Bar, salt-marshes have developed in the shelter of the shingle structures, which as one might expect are sandy and resemble west coast sites in Britain rather than the more silty marshes of Norfolk.

In complete contrast Barra lies at the southern end of the chain of islands known as the Outer Hebrides. The islands are composed mainly of Pre-Cambrian Gneiss, with some granites and gabbros. These essentially resistant rocks form only relatively low relief in Barra, and North and South Uist, limited mountainous terrain being confined to Harris and Lewis. Everywhere the coastline is fretted by a myriad of sea inlets, with numerous islets and sea stacks. Long-continued erosion, especially by glacial action, has been responsible for severe scouring of the rock surface (cf. photograph 32). Ice moved north-westwards from the Scottish mainland to coalesce with ice masses forming over the island archipelago during the Devensian glaciation. The present day scenery results chiefly from the post-glacial (Flandrian) marine submergence of this ice-scoured area, creating a coastal landscape in Barra of islands, reefs and skerries, some linked by plugs of glacial till, but many more joined by the enormous accumulations of sand and the comminuted marine shells that form the bulk of the machair. Drowned peats and stumps of forest trees testify to recent and continued marine submergence, for

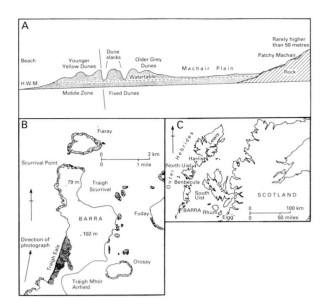

Barra, Outer Hebrides
A. Schematic diagram to indicate the normal relationship between coastal dune systems and the machair.
B. Coastal features and photograph location.
C. Geographical location of Barra.

Barra, together with the whole of the Outer Hebrides lies within a zone where the rate of rise of sea level still averages about 15 cm per century. Because the islands appear not to have experienced substantial isostatic recovery following the last glaciation, elevated shorelines and relic cliffs and sea caves, such as those recorded in the Inner Hebrides and on the Ayrshire coast, are absent. Elevated post-glacial shorelines are found only at low levels (+2 to +5 m OD). In an area now considered to be virtually stable so far as isostatic movement is concerned, and in the presence of slowly rising sea level, encroachment and erosion by high energy wave action is especially severe along the exposed western coasts of the islands.

This is a landscape of mobility and change, as photograph 109 illustrates, with considerable sand drifting from west to east, from Traigh Eais across the isthmus linking two rocky ice-scoured promontories towards Traigh Mhor. Significant sand blowing can be expected in this exposed windy climate, and such landscapes are repeated in many coastal areas of the Highlands and Islands. Some of the dune–machair systems are believed to date from about 6,000 to 2,000 years BP, but on Barra the dunes are still mobile and invading the machair pasture, while in places wind erosion is cutting deep, narrow blowouts from west to east in the higher machair surface. These machair surfaces or 'links' occur with the coastal sand dunes in some 60 per cent of the beach–dune systems along these coastlands of north-west Scotland. The coastal dunes give way to a smooth or slightly hummocky surface with a grass sward typically of red fescue, meadow grass and species of clover and plantain replacing the characteristic dune plants such as marram grass, sand sedge (*Carex arenaria*) and sand couch grass. The mature grassland of the machair forms a distinct ecological–topographical unit, where typically lime-tolerant species of grasses grow on calcareous shell sand derived from the comminuted remains of sea shells. Where the sand lacks a shell content and consequently is siliceous, then acid conditions prevail, heathland appears and the true machair becomes patchy, as shown on the diagram above.

In the Outer Hebrides, it has been calculated that 55 per cent of the beaches by area have between 41 per cent and 70 per cent lime content and consequently the extent of the machair is considerable and it seems likely that the machair systems are composed of a complex of surfaces representing repetitive phases of deposition and erosion. These phases have been recognised and dated by examination of the buried soil horizons and found to contain archaeological material: many of the machair surfaces are believed to be up to 4,000 years old. However, landscapes such as that depicted in the photograph, and which are repeated throughout the Highlands and Islands, represent mobile beach-dune-machair landforms with much shorter time-scales. These range from a few hundred years for stable machair surfaces to less than a single year for the active dune forms according to the work of A. S. Mather and W. Ritchie.[17]

Salt-marshes and bare mudbanks characterise Langstone Harbour, which forms the central portion of the indented coastline between Gosport and Chichester, with Chichester Harbour and Portsmouth Harbour lying respectively to the east and west. These harbours represent the drowned portions of river valleys which had been entrenched in the low coastal plain during the period of low sea level in the late Pleistocene. At that time the coast was several kilometres to the south of its present position and the streams draining the chalk dipslope of the South Downs flowed southwards in broad, shallow valleys cut in the Tertiary clays, gravels and brickearths of the coastal plain. The encroaching Flandrian sea drowned part of the former coastal plain and initiated the complex of inlets we see today. Deep alluvial muds up to 9 m deep overlie the solid geology as a result of limited aggradation from some freshwater streams, while the bulk of these lime-rich clay silts are of marine origin. The harbour shores show evidence of both aggradation and erosion, and there has been considerable reclamation near Portsmouth and Havant.

109 Barra, Outer Hebrides. Here we see the Eoligarry Peninsula, a small area at the northern extremity of Barra, where between the rocky outcrops of Ben Vaslain and Ben Eoligarry, formed of tough igneous and metamorphic rocks, there is a highly mobile landscape of drifting sand extending inland from the exposed western (Atlantic) beach of Traigh Eais, on the left. This is an example of a beach-dune-machair system. An area of coarse shell-sand, where large blowouts dominate the sand dune area, can be seen to the left of the photograph. The Eoligarry Machair occupies much of the remaining vegetated area, but several deep, narrow blowouts are active and have deeply dissected part of the machair surface in the foreground. NF 702079.

Langstone Harbour, Hampshire
The location of the vertical air photographs 110a,b,c are shown in Langstone Harbour, which is enclosed by Hayling Island and Portsea Island (now the peninsula between Cosham and Portsmouth). The changing vegetation cover (*Spartina*) of the inter-tidal mudbanks at Site C is illustrated for the period 1946 to 1980.

There is very little freshwater entering Langstone Harbour at present and the saline mudflats and marshes experience inundation twice daily under the tidal regime. Numerous gravel banks are found within the harbour and the entrance is restricted by substantial accumulations of angular gravel (now shingle) between Portsmouth and Hayling Island. The maximum tidal range within the harbour is about 4 m and within this protected environment substantial salt-marshes have developed. On the original bare areas of bedrock, sandflats, and more particularly the inter-tidal mudflats, vegetation successions have grown as colonisation by various salt-tolerant plants has taken place. The pioneer plants such as glasswort (*Salicornia herbacea*) require protection from wave action. Glasswort seeds require three days undisturbed by waves or tidal currents to enable their roots to become established in the soft mud and clearly only salt-tolerant species can begin the process of establishing a vegetation sward across tidal flats where salinity may reach 16 per cent during the warm summer months. The establishment of a stable substratum is aided by the growth of green algae while glasswort can aid accretion of sediment, at recorded rates of up to 3 cm/year, sea grass (*Puccinellia*) up to 10 cm/year and cord grass (*Spartina*) up to 5–10 cm/year. Thus marshes can grow both upwards and outwards as more sediment is trapped, and as successions of vegetation are established according to the tidal regime, salinity and other factors. The ebb and flow of the tides across the tidal flats allows deposition of sediment, which is encouraged as vegetation spreads and stabilises the marsh surface. The continuation of this process does not produce equal amounts of deposition over the whole area and thus irregularities of accumulation lead to the development of low hummocks and shallow depressions. The latter can develop into drainage channels, which in time become part of a well-developed creek system, and in due course the marsh continues to grow, and new plants appear to colonise the creek and marsh surfaces. As the marsh gains in height the tidal waters submerge the marsh surfaces for shorter periods and the ebb and flood tidal flows move along the creeks and spill occasionally across the marsh surface, and in turn new plants will colonise the rising surface. A certain amount of erosion can obviously occur along the creeks, especially as ebb flow takes place, and as the marsh extends upwards and seawards micro-cliffs can develop in the outer face of the marsh. Erosion and deposition can therefore occur in close proximity on a salt marsh, and this is true in Langstone Harbour where cord grass has introduced a somewhat different pattern of marsh development. Hybrid species of cord grass began to invade Britain's southern coasts shortly after 1900. The large inter-tidal area of Langstone Harbour became dominated by the hybrid (*Spartina anglica*) and the usual general mixture of salt marsh plants virtually disappeared. However, during the last 15 years or so cord grass has been dying back and eel grass (*Zostera marina*) extending on the low-lying mudflat below mean tide level. The photographs (110a, b, c) show the different types of cord grass marsh which have developed, while the accompanying maps also illustrate the changes that have taken place over time. The decline recorded for cord grass and the subsequent expansion of eel grass implies the replacement of areas of high-level marsh and the redistribution within or complete loss of large volumes of marine silt from the harbour. Within the harbour, short wavelength, high amplitude, high-energy waves can be generated during storm

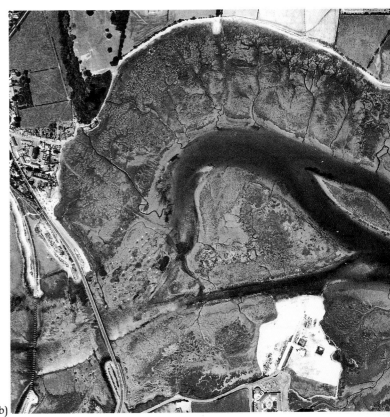

(a)

(b)

110a, b, c Langstone Harbour, Hampshire. These vertical photographs depict the variety of salt marsh found within Langstone Harbour, where the dominant vegetation of the inter-tidal mudflats is *Spartina anglica*. The sites chosen illustrate the presence of mature marsh with healthy *Spartina* (a), mature marsh with channels and some mounds bare of vegetation, (b) essentially bare inter-tidal banks with incipient creek systems and very small patches of *Spartina* (c). At this latter site a comparison has been made to show the changing density of *Spartina* cover between 1946 and 1980 (see the map on p. 215). SZ 686995.

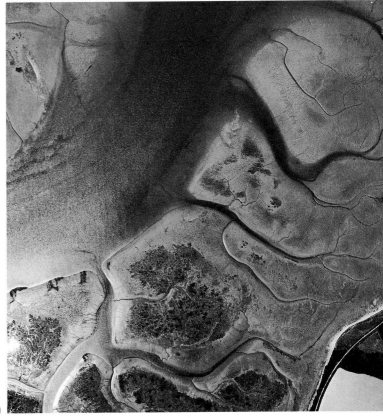

(c)

conditions. These waves can cause considerable erosion, prevent accretion of sediment and new growth on the mudflats, and thus have contributed to the overall decline of the cord grass cover.

The several different aspects of the distribution of cord grass can be identified in different parts of the harbour. At *site A* there is a mixed marsh association, including cord grass in a healthy condition and rooted in fine muds. The marsh is, however, dissected by a dendritic system of creeks and the mature marsh ends quite abruptly in a micro-cliff, clearly indicating that considerable erosion has already occurred.

At *site B* mature cord grass marsh occurs at the head of Emsworth Harbour but a reticulate pattern of channels has developed, and close by on the foreshore at Langstone Village the marsh surface has broken up into isolated mounds. Some of these mounds are bare of healthy cord grass, but mud hummocks remain because of the binding effect of the rhizomes of the dead plants. There is some vigorous cord grass growth in places on the new, lower marsh surface.

At *site C* in the south-eastern part of the harbour we can observe bare intertidal mud banks, with incipient creek systems, and some patches and narrow strips of *Spartina* adjacent to the main channels. Comparison of the maps of this area for 1946 and 1980 indicates the extent of the 'die-back' of the *Spartina* in recent decades, and the presence of micro-cliffs and isolated *Spartina*-carrying islets in the mudbanks also shows that erosion and retreat of the marsh has been in progress for some 40 years. Such 'die-back', without the establishment of a new low-level marsh over most of the harbour, may well allow further erosion, and loss of sediment, but additional studies are necessary to establish long-term trends in the ecosystem of the harbour.[18]

CLIFFED COASTS

There exists in Britain a wide variety of cliffed coasts, their form depending to a considerable degree upon the rock types and structures, and the relief of the land in the immediate vicinity of the coast (see the map on p. 220). Exposure to wave attack and longshore movement of sand and shingle may influence locally the form of the cliff profiles and determine whether or not a beach or shore platform is present that can contribute to the damping down of wave action and the protection of the base of the cliffs. Other processes, such as sub-aerial weathering of the rocks, and mass wasting of coastal slopes, for example by landslips and massive rock falls, can be important factors in determining cliff morphology. As we have already seen there is also an inheritance factor, for some segments of coastline reflect the action of processes over hundreds of thousands of years, spanning glacial and interglacial periods. Where hard-rock cliffs are present, extending up to several hundred metres above sea level, the time-scale for development is likely to be much greater.

In North Devon, between Ilfracombe and Foreland Point, Lower and Middle Devonian rocks are exposed. The east to west trending coastline cuts across the strike of the main geological structures and intersects a 'stepped' landscape which is deeply dissected by a number of valleys containing extremely powerful streams draining the northern part of Exmoor. The main watershed of these streams lies

111 *opposite* Ilfracombe–Lynmouth Coast, North Devon. On this stretch of coast between Lee Bay and Foreland Point, the coastal slope from the Exmoor plateau is steeply inclined towards the Bristol Channel. The Valley of the Rocks and the col at Lee Abbey represent segments of former fluvial valley systems, now abandoned, and perched some 80–100 m above sea level. The coastline thus consists of a sequence of vertical cliffs and extensive sub-aerial slopes interrupted by the abandoned 'dry' channel of the Valley of the Rocks and the Lee Abbey Col. Wringcliff Bay and Lee Bay represent two small 'pocket' bays with shingle beaches along a coastline otherwise dominated by rocky cliffs and reefs (see maps on p. 220). SS 705495.

only between three and eight kilometres from the coast. The highest and finest cliffs are found between Combe Martin and Foreland Point, where extensive sections exceed 200 m in height, for example, at Great Hangman, and the cliffs below Holdstone Down and Trentishoe Down. The morphology of the cliffs is shown in photograph 111 and the accompanying map, where hog's back profiles are displayed.[19] These cliffs represent the limited truncation of the smooth, rounded slopes forming the northern edge of the Exmoor Plateau, and at many places the vertical extent of the cliffs is very restricted. Extensive marine abrasion platforms are generally absent at the base of these cliffs, and the few small 'pocket' beaches are boulder strewn, and backed by steep rock faces, or in some cases by cliffs developed partly in periglacial head deposits (e.g. at Lee Bay). The hog's back forms become less impressive and decline in height west of Little Hangman, and the coastline consists in part of small rocky promontories and inlets, such as Watermouth Bay, where the sea has breached the northern side of an old river valley.

The Valley of the Rocks and the col at Lee Abbey are believed to represent part of an abandoned valley system now truncated and dismembered by marine erosion. Elements of the north-flowing drainage system off the Exmoor plateau to the Bristol Channel are considered to have been 'captured' by progressive marine attack and coastline recession: the East Lyn river and its tributaries have thus been 'captured' at Lynmouth, having formerly flowed westwards to Lee Abbey, Lee Bay and beyond.[20] This hypothesis assumes that wave attack has been able to exploit weaknesses in the rock strata and remove enormous volumes of rock along this high, cliffed coastline. However, in places, the presence of the hog's back cliffs suggests only limited marine erosion, supported by the restricted development of vertical marine cliffs and the absence of wave-out abra-

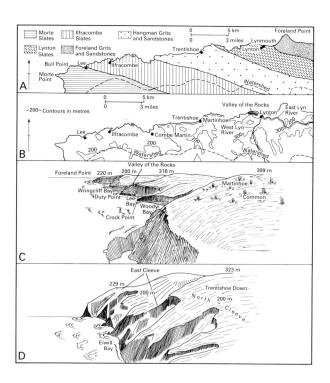

North Devon coast between Ilfracombe and Foreland Point
A. Geological map (after Geological Survey).
B. Relief Map (based upon the Ordnance Survey).
C. The coastal slope between Trentishoe Down and Foreland Point, where both vertical and hog's back cliff profiles can be seen, together with the Valley of the Rocks.

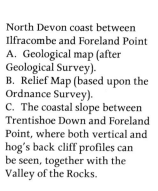

sion platforms at the base of the cliffs. An alternative explanation suggests that the elevated dry valleys and cols represent remnants of meltwater channels, which evolved when ice occupied the whole of the Bristol Channel. Such meltwater streams (marginal or subglacial) may have developed, or enlarged, channels which we now see as dry valleys and isolated cols. The evolution of the detailed morphology of this segment of coastline is thus still the subject for conjecture and future research, and must be related in part to the development of the Bristol Channel.

The Gower Peninsula projects boldly into the Bristol Channel and it is mostly built of Carboniferous Limestone. While the cliffs seldom exceed 61 m in height and the coast cannot match the grandeur of the Great Orme cliffs in North Wales, the Gower cliffs are of great physiographic interest. A series of low plateau surfaces fringes the Gower coast (e.g. at 61 m and 122 m OD), of which the 61 m level is well displayed in each of these photographs. Swansea Bay and the Burry River estuary, which define the eastern and western limits of the peninsula are located in the relatively softer Upper Carboniferous rocks, while synclines in the shales of the Millstone Grit series have been eroded to form major bays at Oxwich and Port Eynon.

The Carboniferous Limestone and Millstone Grit Shales are arranged in a series of tight Armorican folds trending west-north-west to east-south-east in an *en échelon* pattern from Mumbles Head to Worms Head, where the most westerly of the anticlinal structures is exposed in the cliffs and the narrow spine of the striking promontory. In photograph 112a the 61 m level can be seen to bevel the synclinal and anticlinal structures in the Carboniferous Limestone between Worms Head and Port Eynon, the north-east dipping strata of the southern limb of the syncline giving rise to steep, craggy cliffs.

The anticlinal axis passes between Kitchen Corner headland and the Inner Head, with the limestone dipping on average at angles of 80° to the south, and with similar high dips northwards in the fretted cliffs extending to Tears Point. Along this section of coast nearly every inlet, gully, and cave-system is eroded along north-east–south-west trending fault lines or joint systems in the limestone. In photograph 112b, a typical 'blind dry valley or 'slade', known as The Knave, breaks the continuity of the strong cliff line. Solutional processes have also influenced the micro-topography of the cliffs, and especially of the shore platforms at their foot, which in photographs 112a and b are seen to be mainly intertidal features. The lower parts of the shore platforms are deeply dissected by solution hollows, with intervening pinnacles, although at the cliff-foot the actual 'platform' may be nearly horizontal, with occasionally displaced joint blocks. Worms Head is also interesting for Pleistocene geomorphologists because small outcrops of mixed lithology drift occur, sealing either a red clay (or soil?) of possible interglacial status, or raised beach deposits of possible Ipswichian or Last Interglacial age.[21]

The continuity of the plateau-like nature of Gower is illustrated in photograph 112c, the 61 m surface truncating the folded and highly tilted limestone at the coast, while inland gentle gradients lead back to the 122 m level. These low surfaces may represent wave-planated features – the generally preferred explanation – of very considerable age, and certainly of pre-Pleistocene age. At the centre

(a)

(b)

(c)

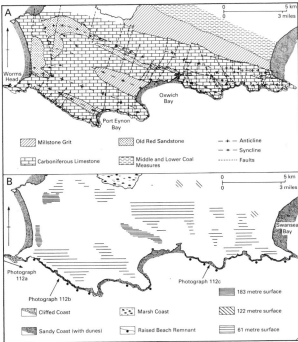

112a, b, c *above and opposite* The Gower Coast, South Wales. Three photographs illustrate the variety of coastal topography in the Gower Peninsula, where the cliff forms and details of the shoreline morphology are dependent to a considerable degree on the macro- and micro-structures of the Carboniferous Limestone bedrock. Photograph (a) shows the islets or large sea stacks of Worm's Head at the south-west corner of the peninsula; (b) and (c) provide contrasting views of the southern coastline. A. SS 383877; B. SS 432863; C. SS 563869.

Gower, South Wales
A. Geological map of the peninsula.
B. Coastal geomorphology and photograph locations.

of this photograph is Hunt's Bay (or Deep Slade), a fault- and joint-controlled blind valley. In contrast to The Knave, however, the rock floor is obscured by many metres of superficial deposits: the latter form the prominent cliff 4–6 m high at the rear of the inter-tidal wave-cut platform (cf. photograph 69). Much of this superficial cover, which extends to rocky screes lying against the cliffs of the embayment, has been produced under periglacial conditions and moved downslope to form an apron of debris which has been cut back by the sea, and is still being eroded at the present cliffline. The inter-tidal platform exposed in the foreground carries cemented patches of the so-called 'Patella' or Last Interglacial beach and the rock surface appears to pass without significant break below the superficial deposits. These extensive wave-cut platforms are probably composite features in terms of age, and represent wave activity and solution processes acting over long periods of Pleistocene and Holocene time.

The Dorset coast extending from Durdle Door in the foreground to Worbarrow Bay in the distance displays a considerable variety of cliff forms and a fine example of the differential erosion of a series of Jurassic and Cretaceous strata (photograph 113). The steeply dipping Jurassic Limestone (Portland beds) from an outer rim of nearly vertical cliffs, which were once continuous from lower left to upper

113 Lulworth Cove and Worbarrow Bay, Dorset. This segment of coastline provides a good example of differential erosion by the sea of a variety of bedrock types. The breaching of the more resistant outer rim of Jurassic rocks has led to the development of inlets of various size, ranging from Stair Hole in the immediate foreground, to Lulworth Cove in the middle distance, and to the much larger Worbarrow Bay. The sea is now in contact with the chalk downlands over substantial sections of coastline, where active recession of the cliffs is taking place. SY 826797.

right across the photograph. This more resistant limestone (and the Purbeck Beds) has been breached by the sea in a number of places as shown in the map.[22]

The smallest breach is at Stair Hole where a small inlet allowed the sea fresh access to softer Cretaceous strata (Greensand, Gault and Wealden Clay), but while land-slipping is steadily aiding the enlargement of this new bay, the sea has not yet reached the Chalk, which forms the coastal downlands. Lulworth Cove represents the second stage of evolution of the coastal scenery, the sea having enlarged an initial arch and breach in the outer Portland Limestone cliffs and, aided by weathering and mass movement, has opened out a horseshoe-shaped bay, which extends into the chalk downlands. Worbarrow Bay, in the distance, represents the third stage of modification, with complete loss of the protective outer limestone cliffs over a distance of some two and a half kilometres between the Mupe Rocks and Worbarrow Tout (or Headland).

The varying height and morphology of the marine cliffs thus reflects the alternation of relatively strong and weak rock types and the intersection of a prominent chalk ridge rising to over 120 m. The coastal evolution appears also to have been influenced by the presence of valley systems leading into Lulworth and Worbarrow Bays, which can be seen in the photograph. These valleys have contributed to the breaching of the chalk ridge, the excavation of the softer Greensand, Gault and Wealden Clay beds, and probably also to the breaching of the Portland and Purbeck Limestone series, although this is less clear. Thus, while marine erosion is the most important factor at the present time, the coastal forms certainly owe some aspects of their morphology to the invasion and drowning of pre-existing river valleys.

Where the South Downs are intersected by the present coastline between Seaford and Eastbourne in East Sussex, the Chalk forms a line of precipitous cliffs which at Beachy Head, reach nearly 160 m in height (photograph 114). Wave action during the last 5,000 to 6,000 years has cut into the downlands to such an extent that nearby a series of dry valleys 'hang' above the beach and the rounded summits of the watersheds form the well-known Seven Sisters.

The massive headland of Beachy Head consists of vertical faces developed in the well-jointed chalk, which in places are obscured by talus cones, and the exposed beach is littered with blocks, derived from frequent rockfalls, resting upon a wave-cut abrasion platform, on which the lighthouse stands. Despite the considerable width of this inter-tidal platform, representing retreat of the cliffs, and the removal of large volumes of rock, the rate of consumption of the cliffs is also closely related to height of the cliffs and the inherent instability of the chalk bedrock.[23] The large talus cone extending to about half the height of the cliff testifies to recent massive slope failure. The extension of the inter-tidal platform and retreat of the cliffline also depends on the rate at which mechanisms such as the solution of the chalk debris takes place, as well as development of pore water pressure in tension and joint cracks and the action of freeze–thaw processes on the chalk bedrock. Some large falls in chalk cliffs also exhibit evidence of failure of the rock by shearing, most often related to smooth, steeply inclined (up to 65° has been quoted) major joints, although a series of closely-spaced minor joints may also provide shear planes. The rate of retreat of chalk cliffs near Beachy Head has been quoted as between 0.5 and 1.25 m per year,

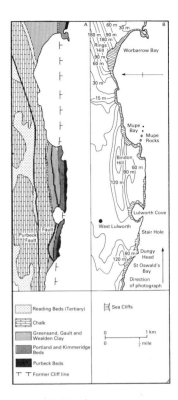

Lulworth–Worbarrow, Dorset
A. Generalised geology of the Dorset Coast from St Oswald's Bay to Worbarrow Bay.
B. Coastal morphology.

114 Beachy Head, East Sussex. Some of the highest cliffs developed in Cretaceous Chalk bedrock are shown here, towering nearly 160 m above the present shoreline. Considerable segments of the cliffs have vertical profiles, but the large talus cones shown here indicate that there is continued instability of the chalk. TV 583952.

Beachy Head, Sussex
Aspects of the topography of the coastal zone between Seaford and Eastbourne.

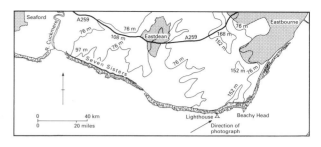

but average figures probably conceal the importance of catastrophic falls of the type seen in the photograph.

Many kilometres of coastline in southern England display forms of instability, of which the most spectacular are major multiple rotational landslips, usually involving Chalk or Upper Greensand as a caprock, a stiff over-consolidated clay such as the Gault, and a massive substratum.[24] Examples of such landslips have been described from Folkestone Warren in Kent,[25] along the southern coast of the Isle of Wight and near Lyme Regis (see chapter 7). Associated with these massive failures of the rock strata are zones of mudsliding, where mudflows contribute to continued degradation of the coastal slope : such mudslides also characterise the cliffs developed in London Clay around the Thames Estuary, where shallow rotational landslips also contribute to recession of the cliffs.[26]

In contrast, the cliff forms developed on resistant rocks in northern Britain also display considerable variety, as already illustrated in photographs 16, 24, 32, 35, 101 and 102. Two more examples are selected from the Orkney Islands, the nearest group of islands to the northern tip of mainland Scotland (Caithness), and separated by the storm-ridden Pentland Firth.[27] This archipelago consists of some 70 islands and numerous rocky skerries, set in a complex pattern of land and sea-lochs and firths. The sea channels separating the islands probably represent the final stages in the drowning of a rocky landscape where former river valleys and ice-gouged lowlands have been submerged by the post-glacial (Flandrian) rise of sea level. Most of Hoy consists of coarse, pebbly sandstone rocks of Upper Old Red Sandstone age, rising above a basal volcanic series of lavas. The very exposed west-facing coast consists of magnificent sea cliffs sometimes exceeding 300 m in height. The sea stack known as the Old Man of Hoy rises as a substantial pillar of horizontally-bedded sandstone some 137 m above the

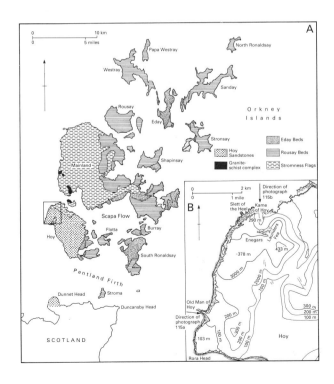

Orkney, Hoy
A. Generalised geology of the Orkney Islands.
B. Coastal topography of north-west Hoy.

(a)

(b)

volcanic lava forming a basal wave-cut platform at the foot of these cliffs (photograph 115a). Even these tough sandstones continue to crumble as the processes of wind, rain, wetting and drying of the rock, frost action and chemical decay operate on the vertical faces. The very existence of the sea stack testifies to the combined effect of these sub-aerial processes and the ability of the powerful Atlantic waves to remove much of fallen material from the cliff-foot. It seems improbable that such a sea stack could survive any over-running of the area by ice during the Devensian glaciation, and consequently this may be an indicator, albeit a crude one, of the rate of retreat of these cliffs over the last 14,000 years or so.

However, at the northern end of the island we can observe (photograph 115b) that substantial modification of the cliff profile has taken place at Kame of Hoy. Here, a local glacier excavated a large cirque, some 200 m deep in the plateau edge. The lip of the cirque is notched by a deep gully and 'hangs' about 100 m above the sea. The sea cliffs are still imposing and retain vertical profiles below the lip of the cirque, but give way to scree-clad slopes to the right in the photograph.

Considerable stretches of the coastline of eastern England are composed of relatively unresistant glacial deposits, where erosion has been severe because of the inherent weakness of the glacial till and fluvioglacial sand and gravels, together with post-glacial alluvium. Comparatively rapid recession of the cliffline is still occurring in parts of Norfolk and Suffolk, while in Holderness (Yorkshire) dramatic changes have taken place since Roman times (see the map below), and there have been significant losses since the Domesday Survey in the eleventh century.

However, along the 60 km of the Holderness coast it has been shown that there is marked variation in the rate of erosion, which is largely controlled by

115a, b *opposite* Hoy, Orkney Islands. The sea cliffs on the northern and western coast of Hoy are formed in horizontally-bedded Old Red Sandstone bedrock. The sea stack known as the 'Old Man of Hoy' (137 m high) is seen in photograph (a) against a background of nearly vertical cliffs, which nearby at St John's Head rise to 348 m. Screes testify to the continued retreat of these cliffs, amongst the highest vertical sea cliffs in Britain. In (b) a low-level glacial cirque is seen to break the continuity of the cliffs at the Kame of Hoy. HY 202047.

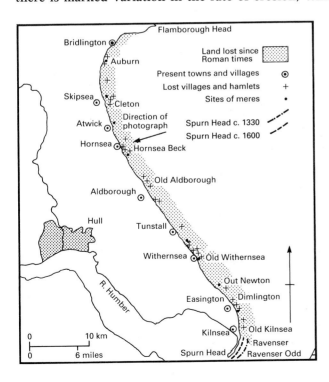

Holderness, Yorkshire
The rate of erosion of the Holderness coast between Bridlington and Spurn Head is illustrated. The position of the Roman coastline is approximate, but documentary evidence indicates losses of land averaging about 2 m per year, and the loss of a strip of land about 4 km wide since Roman times, in cliffs up to 30 m high in glacial drift.

229

the presence or absence of an upper beach with a marked convex profile. Where this upper beach is reduced in height, or entirely absent, wave energy can be expended directly against the cliff-foot at neap as well as at spring tides. A water-filled channel, or runnel, can lead towards the cliff and a wave-planated shore platform is usually present, eroded in glacial till. This type of beach feature is known as an 'ord', and is described as being 1 to 2 km in length on average, and there are usually ten 'ords' present along some 60 km of coast (see the diagram on p. 231). Over time the 'ords' migrate southwards and thus bring successive segments of the coastline under severe wave attack. The presence of an 'ord' can

116 Hornsea Holderness, Yorkshire. The coast between Flamborough Head and Spurn Head is developed largely in glacial deposits, with cliffs varying in height, here less than 15 m, and protected by sea walls and with systems of groynes to assist beach stabilization. The general drift of beach material is from north to south (right to left in the photographs), as is shown by the zig-zag pattern of beach accumulation against the groynes. Where groynes such as these are established then other segments of coastline can be starved of beach material and erosion may occur. Hornsea Mere is the last surviving example of a series of former salt-water creeks which occupied shallow valleys in the glacial deposits, or were depressions containing freshwater in the hummocky drift terrain of Holderness. Most meres have disappeared as a result of erosion by the sea or silting-up and artificial drainage. TA 205478.

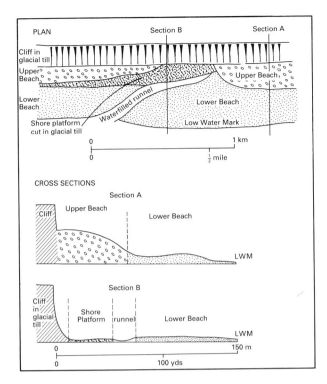

Holderness, Yorkshire
A generalised plan-view of a typical 'ord', together with two cross-sections to illustrate the shoreline conditions; it is apparent that with an 'ord' in position the water-filled runnel and absence of the upper beach can allow larger waves to reach the cliffline under both neap and spring tides: as a result increased erosion can take place of the cliff, and groyne systems and seawalls can be undermined if the foundations are not sufficiently protected.

bring about considerable damage to seawalls, and groyne systems of the type featured in photograph 116[28] because of increased wave energy being expended in the deeper water associated with the runnel and the absence of the upper beach. Interference with the natural movement of beach material by groynes (photograph 116) necessarily starves other parts of the shoreline of the sand and shingle required to maintain stable beaches.

Substantial coast defence works have been carried out along sections of Britain's east coast, but an additional problem along the North Sea coast has been caused by storm surges and the accompanying exceptional wave activity recorded during the 1953 and 1978 sea floods, a topic to be considered in more detail in chapter 7.

GUIDE TO FURTHER READING

E. C. F. Bird and M. L. Schwartz (eds.), *The World's Coastline*, New York, 1985.

C. A. M. King, *Beaches and Coasts*, London, 1959.

J. Pethick, *An Introduction to Coastal Geomorphology*, London, 1984.

A. Robinson and R. Millward, *The Shell Book of the British Coast*, Newton Abbot, 1983.

J. A. Steers, *The Coastline of England and Wales*, Cambridge, 1946.

J. A. Steers, *The Sea Coast*, London, 1954.

J. A. Steers, *The Coastline of Scotland*, Cambridge, 1973.

7   Changing landscapes

INTRODUCTION

The title of this chapter can be seen to be directly related to one of the main definitions of geomorphology, the study of the evolution of landforms. In this sense, the aspect of time is introduced as it was at the turn of the century by W. M. Davis in his publication which established a positive relationship between geological structure and lithology, geomorphological processes and the stage of evolution which a particular landscape had achieved.[1] Despite considerable critical discussion of Davis' views the modern development of British geomorphology has proceeded from straightforward landform description and denudation chronology through detailed process studies, often dealing with the drainage basin as a fundamental landscape unit, to the modern era of applied studies.[2] However, time has still remained a fundamental element in approaches toward landform studies, and we find for example, that Gregory[3] places morphological elements (F), processes operating in the physical environment (P), and the materials (M), upon which the processes operate over changing time periods (dt), into a general equation:

$$F = f(P, M)\, dt$$

At first sight, this expression would seem not to have moved very far from Davis' triology of 'structure, process and stage', but as Gregory explains,[4] the equation encompasses a number of different approaches to studies in physical geography. Firstly, the study of the elements or components of the equation – landform, climate or geomorphological processes – in their own right, either qualitatively or quantitatively. Secondly, the study of the way in which the equation operates at differing levels, for example mega- or microscale geomorphology. Thirdly, the analysis of the way in which the equation varies over time with respect to changing environmental constraints. Gregory comments that it is at this stage that the importance of human activity and its interaction with environmental systems has to be considered. Fourthly, there is the application of the results of the study of the equation, involving the extrapolation of past trends in order to predict future ones.

This chapter has developed out of the above considerations and attempts to concentrate on applied aspects of geomorphology in relation to changing landscapes. Although changes resulting from the operation of geomorphological processes can be discerned, quite often the change is subtle or indistinct even to a trained eye. Thus the photographs do not relate to a specific site or landform viewed over time. Rather, they relate to more obvious, and perhaps highly significant, elements of landform evolution such as periodic river and coastal flooding, catastrophic landslides and important man-induced changes.

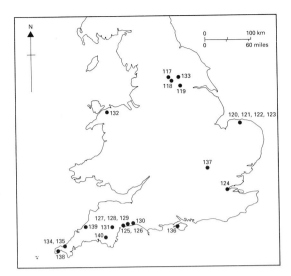

Location of air photographs.

Thus while many of the photographic plates indicate the essential dynamic nature of geomorphology they cannot encompass the exact magnitude or frequency of landform change whether by climatic or human control. Quite often landscapes seem to evolve in a 'passive' manner and observing or studying 'active' phases of landform evolution can be inherently dangerous. We are therefore lucky if the results of such active phases are caught on film. Consequently much geomorphological work goes unseen or even unknown, and as a result our knowledge of changing landscapes and the processes controlling them is still somewhat sparse.

The selection of suitable illustrations has been governed both by the material available in the Cambridge Air Photo Collection and in the attempt in this concluding chapter to complement the presentations of the other authors, while avoiding any substantial overlap with the air photographs in chapters 1–6.

FLOODS AND FLOODING

Floods can result from a variety of causes. Ward identified the following main groups: climatological (e.g. rain and snowmelt), part-climatological (e.g. coastal storm surges) and a third group which included the effects of landslides and earthquakes or the failure of dams and other man-made structures.[5]

The photographs chosen for this section illustrate combinations of climatological and part-climatological situations. Photographs 117 to 119 show the effects of snowmelt in the Pennines on the River Wharfe and photographs 120 to 124 relate to the generation of storm surges in the North Sea and their effect on the low-lying coast of eastern England.

As was shown by the Flood Studies Report,[6] in the context of drainage basins the character of any particular flood will depend on a number of inter-related variables: the nature of the bedrock and soil, slope and aspect within the catchment, the area of the basin, type of channel network, extent of flood control and river regulation works and antecedent precipitation conditions. Ward termed these factors 'flood intensifying conditions', recognising that most of these variables operated to speed up the flow of water within the catchment, leading to a potentially catastrophic discharge event when bankfull conditions in the river channel are exceeded, usually at a location down valley in a given drainage system. An example is provided in photograph 119 for the River Wharfe at SE 560395, a little way above its confluence with the Yorkshire Ouse. A significant proportion of the worst floods recorded in Britain have a snowmelt contribution.[7] The combination of snowmelt, rapid thawing and rainfall associated with a passing mid-latitude depression, and catchment size can result in serious flood hazards in both upland and lowland sectors of many river systems.

In the case of the River Wharfe, photographs 117 and 118 show the extensive plateau-like nature of the upland catchment of this river. Photograph 117, taken at the confluence of the River Skirfare (Littondale) with the Upper Wharfe at SD 973691, just above the hamlet of Kilnsey encompasses most of the 200 km² Upper Wharfe catchment, which is made up of tracts of limestone and gritstone moorland stretching from Wensleydale in the north and east over to Ribblesdale, beyond Malham in the west. Snow cover on this 600 m (2,000 ft) planation surface is common in most winter seasons. The amount of discharge generated in the valley systems of the Upper Wharfe is a function of antecedent precipitation and ground conditions, the depth of snow and the rate of melting. To this must be added the considerable groundwater contributions from the Carboniferous Limestone bedrock and the possibility of accelerated surface runoff if infiltration from the melting snow pack is precluded by a frozen soil cover.

Rapid snowmelt in Britain is usually the result of an influx of warm moist air. Rain often accompanies the melt, which most commonly occurs under cloudy skies when the contribution of solar radiation to the thawing process is small compared with condensation and convective heat inputs. These latter factors are usually a function of temperature and thus air temperature has been the basis of empirical methods such as the 'degree-day' technique of snowmelt forecasting. Even in lowland areas, meteorological conditions could lead to a potential snowmelt rate of 42 mm/day water equivalent and Jackson[8] suggested this rate could exist in any part of Britain. In the more northern and mountainous parts of the country the 42 mm/day snowmelt rate could 'continue for several days'. It has been recommended that this figure be added to the rainfall profile during any consideration of winter floods.[9]

In view of the small amount of previous work on snowmelt runoff in Britain and the limitations imposed by a general lack of data, snowmelt discharges have been calculated based on the equation:

$$M = K (\theta - \theta_o)$$

Where θ is measured air temperature, θ_o is the temperature at which melting of snow is assumed to start (e.g. 0°C); K is a temperature index or melt factor and M is runoff resulting from snowmelt.[10] Variations in runoff and infiltration are subsumed in the equation. Obviously, the temperature index K can vary both within and between catchments and thus temperature is not a perfect index of snowmelt potential. An improvement on the 'degree day' technique has been suggested where the potential melt rate, M (mm/day) is given by the empirical equation:

$$M = (1.32 + 0.305 \text{ V})\text{T} - 0.236\text{V}\Delta$$

where T is the air temperature (°C); V is the mean windspeed (Knots); and Δ is dewpoint depression (°C).[11] The inclusion of the windspeed factor V emphasised the importance of the turbulent energy exchange at the interface between the snow and the air. The term '$-0.236\text{V}\Delta$' accounted for the reduced potential for snowmelt when the air was unsaturated. Thus the process of snowmelting competes with evaporation in the use of available energy.[12] In addition, a number of other factors would seem to be important in considering melt rate and subsequent flood discharges. Firstly, since temperature alone is an imperfect measure of snowmelt potential, the nature of the melt will vary with different air masses over the catchment. Secondly, temperature values are often extrapolated from a valley or lowland observation station to a mean elevation for the catchment. In upland Britain, as can be seen in photographs 117 and 118, where snow lies 'on the tops' whilst the valleys are largely snow free, such extrapolations can lead to gross errors. Thirdly, snow packs are able to store melt-water until a liquid water retention capacity is reached. Depending on snow density, incoming rainwater may also be stored. Ultimately, the melting pack will release water at considerably higher rates than those predicted from the melt rate alone. Water emanating from the base of the pack will also increase through time since at the beginning

117 Confluence of River Wharfe and River Skirfare, near Kilnsey, North Yorkshire showing the typical terrain of the Upper Wharfe catchment. Moorland rises in steps from wide glaciated valleys – the steps indicating the strong lithological control of differing bands within the Great Scar Limestone. The high moors have a capping of Millstone Grit. Natural drainage is across or on the gritstone and down or through the limestone. Winter snowcover and subsequent rapid snowmelt provides a relatively sudden input to the hydrological system resulting in a rapid rise of river levels and flood discharges. SD 973691, looking N.

118 The River Wharfe above Grassington, North Yorkshire. This view shows the 600 m plantation surface stretching from Great Whernside at 704 m on the eastern horizon across towards Malham in the west. Large spreads of gravel forming mid-channel and point bars above and below Grass Wood can be seen in the centre of the photograph. The thinner snow cover of the valley bottoms as opposed to the moorland tops is quite evident. SD 985647, looking N.

of a period of positive temperatures some energy is used in raising the temperature of the snowpack rather than in melting. Fourthly, observed variations in flood volume are most likely to be caused by differences in the size of the contributing area or the proportion of the catchment experiencing snowmelt. To this may be added a number of drainage basin characteristics which may interact in a complex way to affect the snowpack energy budget; for example, the pattern of bare rock or soil as opposed to vegetated slopes, the extent of forest cover, and aspect variations will be important, with each landscape element having a different albedo as well as different evapotranspiration and temperature characteristics.

Thus the prediction of snowmelt discharge and potential flooding demonstrated by photographs 117 to 119 is still a very inexact science. Much work still remains to be carried out on the meteorological aspects of snowmelt, especially for upland catchments such as the Wharfe. Likewise, investigations are needed into the coincidence of heavy rain and frozen ground conditions which are also often associated with snowmelt events.

In addition to flooding the riparian zone adjacent to the river, the resulting discharges are responsible for much of the geomorphological work undertaken by fluvial events. Vast gravel spreads are incorporated into the floodplain deposits of the River Skirfare (photograph 117). Gravel bars and islands appear in the Wharfe (photograph 118) both above and below Grasswood SD 985655 (the 1.5 km² area of mixed woodland shown middle right of the photograph). Although the position of these deposits is relatively 'fixed', considerable annual modifications take place as high river flows overtop the gravels. The materials themselves are derived largely from glacial till deposits left in these formerly glaciated upland valleys. Downvalley, flood discharges whilst still erosive seem mainly to transport the 'fines' of the particle size distribution and thus gravel bars are absent. The lowland of the Vale of York, characterised by a rolling landscape of glacial deposits, is an area across part of which the Lower Wharfe (photograph 119) meanders, bringing its seasonal input of overbank sedimentation (controlled by rainfall and snowmelt) giving rise to a significant floodplain.

119 The lower River Wharfe near Ryther, Yorkshire showing the valley just above the Wharfe's confluence with the Yorkshire Ouse. Flooding of this nature can be the result of either rapid winter snowmelt in the Upper Wharfe catchment or localised summer thunderstorms. Here water completely covers the floodplain up to the rising ground of a low terrace. SE 560395, looking NW.

Coastal flooding is often the result of a combination of long-, medium- and short-term processes, for example, climatic change bringing about sea level rise, climatic patterns governing the general wave climate and storm surges at a particular coastal sector, and local tidal or meteorological variables affecting local sea level and wave height. The inundation of coastal lands is a serious problem, especially along some of the low-lying areas of eastern England (photographs 120 to 123) where sea flooding has been documented since the thirteenth century. The worst event in recent history was the catastrophic flooding of 1953, when a storm surge occurred during high spring tides, driving water into estuaries already swelled by floodwater draining from inland. In total, over 850 km² were flooded by seawater, 307 people lost their lives and damage was estimated to have cost £30 to 50 million.[13]

In order to fully explain coastal flooding it is necessary to present some detail of the physical processes which are responsible for such inundations. To begin, let us consider the long-term factor of changing sea levels, particularly post-glacial sea level rise. The Quaternary period in Britain was characterised by a number of significant climatic and associated environmental changes. In particular, the take up of oceanic water during glacial advances, the subsequent melting of the ice sheets and the return of water to the world's oceans resulted in sea level fluctuations in excess of 100 m. Such changes in the relatively shallow waters surrounding Britain's coast led to differential loading and unloading of the continental shelf.[14] To this should be added the occurrence of ice masses on land resulting also in isostatic downwarping and recovery, although out of phase with eustatic movements.[15]

There is a general agreement that during the last 15,000 years climatic amelioration has resulted in a release of water from glacial ice causing the rise in sea level known as the Flandrian Transgression. The nature of this post-glacial movement of sea level is interpreted differently by various authors, but two schools of thought have emerged: an oscillatory model based on the work of Fairbridge[16] and a steady rise suggested by Jelgersma.[17] Around stable coasts, allegedly outside the limits of direct glaciation, such as the South West Peninsula, a steady eustatic rise at the rate of 1.5 m per century has been interpreted between 15,000 and 5,000 years BP. The sea level rise then continued at a slower rate to the present day.[18] It is also suggested that a rise of over 2 mm per annum (20 cm per century) is continuing.[19]

It has been recognised[20] that land/sea-level movements will be quite different at various sectors along Britain's coastline depending on the exact rate of isostatic recovery relative to the current overall rise in world sea-level, as shown in the diagram opposite. In general, such upward movements of the land mass are thought to be absent or of little significance in southern Britain as this region was not glaciated during the last main ice advance stage (the Devensian). Thus a picture emerges of a slow continued sea level rise with the increased likelihood of coastal inundation by storm wave activity. On eastern coasts, the situation is a little different. The margins of the North Sea basin are characterised by 'soft' rocks overlain by variable thicknesses of unconsolidated fluvial and glacial deposits. The effect of the Flandrian Transgression has here been exacerbated by localised crustal downwarping at a rate of between 1 and 4 mm per annum.

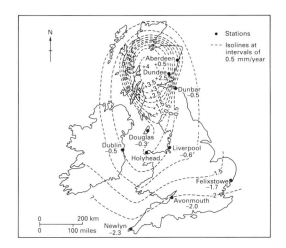

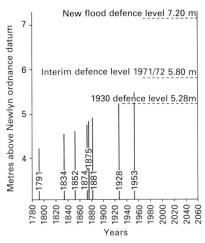

Outline map showing suggested positive and negative movements of the land resulting from the unloading of glacial ice.

Flood tide levels at London Bridge.

Thus the position of high tide mark is increasing in altitude in this coastal zone. In the case of the Thames estuary, and London in particular, archaeological evidence shows that in Roman times the Thames at the site of London Bridge was non-tidal, but by the end of the Roman occupation occasional high tides were starting to enter the Thames.[21] The rise in tide level has continued with records indicating an increase of over 1 m between 1791 and 1953, as illustrated in the diagram above. In fact, the highest recorded tide has reached a level more than 5 m above the river level of early Roman times and there is evidence that this rise is continuing at an increasing rate.[22]

Storm surges add another perspective. Their effect is to convert the normal tidal rise of sea level in the southern North Sea into catastrophic coastal flooding. Such surges are generated by low air pressure associated with the formation and movement of mid-latitude depressions from an Atlantic Ocean source, then tracking east or north-east to the seas of the continental shelf around Britain. The low pressure under the centre of the depression allows sea level to be raised by 10 mm for every millibar drop in air pressure. Some surges are of long duration and may cause an increase in the height of several successive high waters over a period of 24 hours or even longer. However, not all surges are of the same pattern and the peak of a surge may occur at any state of the tide and may not necessarily lead to a dangerous rise above a predicted high water level.[23] Perhaps more importantly, with the passage of the depression, comes the 'drag' factor of gale force winds on the sea surface causing a net forward transport of water. Where this water motion is restricted by the configuration of the land, such as in the 'funnel' of the southern part of the North Sea, a considerable piling-up of water can occur leading to potential inundation. To cause catastrophic coastal flooding, with sea defences overtopped, a storm surge of a given height must combine with a high tide (or at least within an hour or two either side of high water), and the event must occur during spring tides.

Perhaps the best known example of storm surge flooding is the 1953 East Coast Disaster. A deep depression, with a central pressure of 968 millibars – (see diagram on p. 240; similar pressures have been recorded during the 1980s) – with average wind speeds of 144 kilometres per hour and gusts of up to 200 kilometres per hour, passed from the Atlantic on 30 January 1953 to become

The meteorological situation in the early hours of 1 February 1953. The northerly gale force winds, which were associated with the storm surge on the East Coast, are clearly shown.

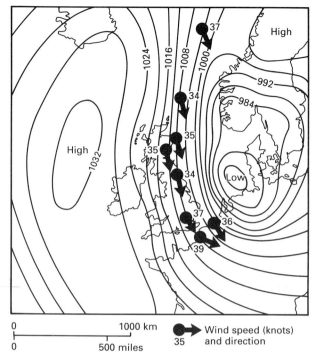

0 — 1000 km
0 — 500 miles

35 — Wind speed (knots) and direction

centred over the North Sea on 1 February.[24] The net result was to build up a bank of water of some $4 \times 10^9 m^3$ in the southern North Sea adding a further 2.5 m height to a predicted spring tide of 3 m above normal.[25] Overtopping of sea defences occurred along the whole of the east coast and into the Thames estuary, causing loss of life, extensive damage to agricultural land and the inundation of oil refineries, cement works, factories, gas works and electricity generating stations, all of which were brought to a standstill.

The events of 1953, although unique in one sense could be repeated. Storm surges are not uncommon as the investigation of 85 surges which occurred at Southend between 1928 and 1938 clearly demonstrated,[26] and a number of high tides have combined with storm surges since that time. A notable event occurred in 1978 as illustrated by the coastal inundation at Wells-next-the-Sea on the north Norfolk coast (photographs 120 to 123).

The low-lying and generally flat nature of the coast in this part of Norfolk is clearly seen in photograph 120, looking south from above the coastguard lookout position at TF 915455. This shows the location of Wells, some 2 km inland of the coast, on a tight bend of a tidal inlet opposite a complex salt marsh area with many small creeks. Behind the coastal dunes (in the foreground of photograph 120) is an area of recreational land use with mobile homes. The dunes obviously provided sufficient protection during the flood event compared with the breaching of the embankment and sea wall along the western side of the inlet. As J. A. Steers[27] commented in relation to the 1953 event, sea wall breaches were mainly formed in most cases from the landward side, after the embankments and sea walls had been overtopped; quite large waves 'inside' the defences demolished walls and outhouses facing the reclaimed marshes.

Clearer views of two breaches at Wells in 1978 can be seen in photographs 121 and 122. At both sites the road linking Wells-next-the-Sea to the coast proper

120 *opposite* Wells-next-the-Sea, Norfolk, after a coastal flood, showing the causeway between Wells and the coastal dunes of the foreground. On the western side of the inlet, behind the road embankment, the salt marsh has been reclaimed. The area is now a mix of fields; nearer the coastal dunes, a sizeable caravan site has been established. The overtopping of the sea wall and embankment along the inlet has spread sand across the caravan site and adjacent fields. TF 915455, looking S.

121 The breached causeway and embankment near Wells-next-the-Sea, showing in detail the nature of the outer breach of photograph 120. The shifting of the embankment and its roadway to the east indicates that the breach was established on the landward side as storm-surge flood water attempted to escape. TF 915451, looking SSE.

was cut due to the inundation, with flooding affecting approximately 50 per cent of the caravan site (photograph 121). The flood water which moved across the site eventually drained away, through the permeable sands of the coastal dunes. What is apparent from both photographs, is that coastal inundation, here causing damage to both property and agricultural land, can include other aspects apart from the flooding itself. The spread of unconsolidated sediments from the breach site across the reclaimed marshes with their fields of crops, clogging up drainage ditches and rendering the soil unworkable until the excess water could be got away and the fields restored, is shown in photograph 122. The accumulation of salt on the surface and within the soil is yet another problem.

122 Sea wall breach near Wells-next-the-Sea, showing the spreading of sand across fields of crops on reclaimed salt marsh. TF 916444, looking E.

The quayside at Wells and the salt marshes on the eastern side of the inlet, which were largely unaffected (as was the case in 1953) by the flood event, are shown in photograph 123. In the town, the 1978 flood surge was sufficiently high to lift a small freighter above the quay wall, leaving it stranded on the top of the quay. It is interesting to note that the surge of 1953 also led to the stranding of a vessel – the 'Terra Nova'.[28] The return period of North Sea storm surges leading to flooding cannot be predicted accurately due to the multiplicity and variability of factors involved. It has been suggested that there is a probability of once in 44 years based on a simple classification of the type of surge and winds

associated with the passage and specific track of a North Sea depression. What is abundantly clear is that there is every likelihood of a repetition of the combination of factors which led to the 1953 flood event at some time, and given the continued negative movement of the land along coastal margins of the southern North Sea there is a serious threat to low lying land in the Thames estuary and even to London itself. Mean tide levels are increasing at Southend by about 300 mm per century and high tide levels at London Bridge by some 700 mm per century.[29]

In London, the 116 km² area at risk from inundation (see map on p. 246) with a population of about 700,000, 250,000 dwellings, numerous government buildings, public utilities and the central part of the underground railway network, has now been protected by a major flood defence barrier (photograph 124) located at Woolwich, some 14 km downstream of London Bridge. The photograph shows the seven supporting buttresses to which the flood gates are attached.

The design of the barrier incorporates a number of criteria:

1 the flood gates, when not in use, were to be positioned below the surface of the river at a depth which did not restrict naviagation;

2 the main seaway openings were to be 61 m wide;

3 the supporting piers were to be as narrow as possible;

4 the reduction in river cross-section limited to 25 per cent;

5 the structure had to withstand a differential head of 9.9 m from an extreme surge.

The construction of the barrage began in 1975, was completed in 1982 and has been described as a considerable feat of civil engineering. With the cost of a major

123 *opposite* The quay at Wells-next-the-Sea. The site of Wells, near the head of a tidal inlet, has often been subjected to coastal flooding usually associated with storm surges. This photograph shows a stranded coaster high and dry on the quay. TF920439, looking W.

124 The Thames flood barrier Woolwich, London. The nine piers of the barrage have four 61 m-wide main navigation openings and two subsidiary 31 m-wide channels. The 61 m width was based on the opening of Tower Bridge, the minimum gap acceptable to the navigation authorities. The curved metal barrier gates are normally kept in the open position – lowered to the river bed. They are raised to a vertical position when dangerously high surge tides are forecast. This rising sector gate design was chosen for its robustness, reliability and simplicity of operation. The completed structure has been built to withstand an extreme storm-surge height of 9.9 m and a reverse head differential of 6.1 m. TQ 415796, looking WNW.

The scale of potential flooding in the London area and the Lower Thames before completion of the barrier at Woolwich.

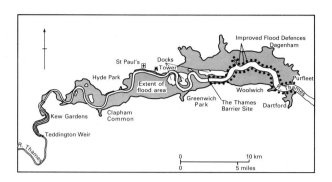

storm surge flood in the London area, estimated at between £3,000 and £4,000 million, simple cost-benefit analysis indicates that such a scheme was justified. In fact, the magnitude of the flood risk to London was in any case unacceptable. In addition, sufficient protection from storm-surge flooding was not provided by the barrier alone. Downstream, on both sides of the Thames, some 112 km of river bank had to be raised and strengthened in order to provide adequate flood defences. Smaller barriers were provided for a number of creeks flowing into the Thames, for example Barking Creek and Dartford Creek, and storm gates installed for the entrances of the Royal and Tilbury Docks. The surge in 1978 brought a flood tide in the estuary to within 0.5 m of the top of the increased height of these sea walls. Since completion, it is anticipated that operational closures to prevent flooding will occur two or three times a year, and by the end of the century the number of closures each year may have increased to ten.

LANDSLIDES AND MUDFLOWS

It has been demonstrated in the previous section on flooding that the physical landscape can often be changed by significant, and sometimes catastrophic, short duration events. The magnitude and frequency relationships of geomorphic processes is still one of the most important factors governing the development of landscape. These aspects are particularly well demonstrated in the highly dynamic coastal zone, especially in those sectors characterised by unstable cliffs experiencing mass movements.

Although landslides can play a general role in coastal recession, where the physical properties of the cliff materials are mixed, with distinct bands of differing lithologies and marked changes in rock permeability, mass movements can be particularly important locally, resulting in very rapid denudation. In such situations, large-scale slope failures often occur due to the combination of undercutting at the base of cliff by wave activity and high pore water pressure in the cliff materials. A whole variety of movements may take place ranging from mudflows, successive rotational slides, slab failures and translational slides. It is often dangerous to generalise, but the failure of thick competent strata may result in massive disrupted blocks and a chaotic undercliff topography, whilst movements involving argillaceous deposits will often be in the form of simple or multiple rotational slides, slab slides and mudflows.[30]

The coastal sector between the mouth of the River Axe (SY 252897) in Devon and St Gabriels (SY 398922) in Dorset contains some of the best examples of landslipping to be found anywhere in Britain (photographs 125 to 130). Apart from their intrinsic scientific interest, the landslips in this area have a direct impact on human activities, as the movements have involved damage to property, the loss of farmland and the destruction of roads.[31] The cliffs which will be considered in detail are those at Culverhole Point, encompassing the Bindon and Dowlands landslips (photographs 125 and 126); Black Ven, to the east of Lyme Regis (photographs 127, 128 and 129); and Cain's Folly and Stonebarrow to the east of Charmouth (photographs 127 and 130). Throughout this coastal section, the cliffs are formed of Lower Jurassic rocks comprising the Lias with some Rhaetic and Keuper Marls (Triassic) west of Pinhay. These are overlain uncomformably by Cretaceous strata which consist of a sandy facies of the Gault Clay, Upper Greensand chert beds and sands, and westwards from Lyme Regis, a capping of Chalk.[32] These Cretaceous rocks are nearly horizontal with only a slight dip to the southeast of 5°, as illustrated in the diagram below.

It is at the base of the Cretaceous strata, in the Foxmould deposits of the Upper Greensand, where the main surface of slipping is located, and where rain water percolating through the overlying permeable strata cannot pass through the largely impermeable Gault Clay. Water accumulates as a perched water table on the clay giving rise to instability and failures in the sandstone rocks above. The overlying chalk (where present) is carried seawards, down-dip as the whole mass slides on a plane of lubrication at the top of the Gault. However, recent work by Pitts and Brunsden[33] suggests that the liquifaction of such a dense material as the Foxmould (Upper Greensand) deposits to form a quicksand is unlikely. In the lower cliff sections, (made up of alternating bands of clays, shales, marls

Sketch map of coastal locations subject to landslipping between Axmouth, Devon and Stonebarrow in Dorset.

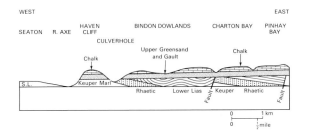

Geological cross-sections of the lithology of the cliffs between Axmouth, Devon and Golden Cap, Dorset.

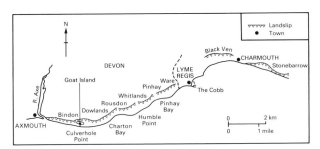

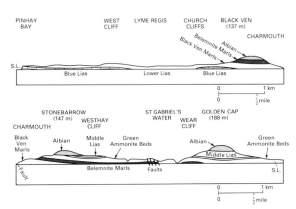

125 The coast between Bindon and Dowlands, Devon, the site of a landslip catastrophe during Christmas, 1839. The area is now a National Nature Reserve of geomorphological, geological and ecological value. SY 290897, looking NW.

and limestones) certain limestones are more resistant than the rest of the Liassic rocks and locally form intermittent water bearing horizons.[34] At Black Ven and on the seaward face of Stonebarrow this lithology has produced notable terraces (see photographs 128 and 130). The clays above these terraces have been washed or squeezed out in the form of mudflows (photograph 129).

The landslipping in the area illustrated has been extensive both spatially and temporally. Pitts,[35] indicated that movements have been occurring since the sixteenth century at Pinhay Bay and details events in 1689, 1828, 1839–40, and 1886 for other areas between Axmouth and Lyme Regis, although the early part of the twentieth century seems to have been largely a period of little or no movement. However, the crest of Stonebarrow to the east of Charmouth began to move in 1942 and slips were also occurring on Black Ven in 1944, while other failures have been ongoing since then, in 1958, 1960, 1966, 1976–77. Undoubtedly, the most significant event was the slip between Bindon and Dowlands, during which some 36 hectares, known now as Goat Island (photographs 125 and 126, and the diagram below) became separated from the main coastal landmass during the Christmas period of 1839. Seaward movement of this mass of land opened a chasm some 64 m deep, extending 800 m from east to west, and some 120 m across at the Bindon end to 60 m at Dowlands. Within the chasm were ridges of chalk and large blocks of land, the surfaces of which tilted backwards towards the main cliff as reversed slopes, and taken as an indication of rotational failure. However, Pitts and Brunsden[36] challenge this view. They point out that as no tilting of Goat Island has occurred, this precludes a rotational failure. Thus the main part of the Bindon landslip is now interpreted as a block slide in the Westbury Shales formation between the Lias and Keuper Marl.

The role of groundwater in landslip activity is clearly important, and in the case of the Bindon/Dowlands slip, at least 100 per cent more rainfall than average

126 Culverhole Point, Bindon/Dowlands landslip, Devon. The detached block of chalk, known as 'Goat Island' separated from the main chalk cliff by 'The Chasm' is clearly seen in the midground of the photograph. Since the 1839 landslip, additional movements have not occurred, although further east along the coast other landslides are recorded. The extensive nature of the site makes this perhaps the most significant landslip in Britain. SY 275893, looking W.

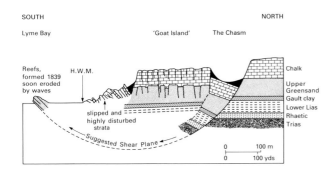

SOUTH   NORTH

Lyme Bay   'Goat Island'   The Chasm

Reefs, formed 1839 soon eroded by waves

H.W.M.

Chalk

Upper Greensand
Gault clay
Lower Lias
Rhaetic
Trias

slipped and highly disturbed strata

Suggested Shear Plane

0   100 m
0   100 yds

The Bindon–Dowlands landslip, Devon; a geological section showing suggested rotational shear plane.

had been experienced immediately prior to the 1839 catastrophe, and since the previous June rainfall had almost been continuous with many gales and associated storm wave activity at the base of the cliffs.[37] A previously accepted hypothesis based on the work of Conybeare, Conybeare and others (cited by Arber)[38] was that the lower levels of the Foxmould sands of the Upper Greensand had been reduced by the exceptionally heavy rain to a quicksand. Thus the whole mass of the Cretaceous rocks overlying the impermeable Liassic strata had slid seawards, throwing up a reef of Upper Greensand offshore some 1,200 m long and standing 12 m above high water level.[39] This reef of rocks was soon eroded by wave activity and the chasm itself is now so overgrown that it is no longer possible to observe the nature of its floor (photograph 126). However, Conybeare's original diagram, with blocks in the base of the chasm tilted backwards, was interpreted as a rotational slide, as the diagram above illustrates.

The pushing up of strata into synclinal forms on the beach slope at or below low water mark, giving rise to low reefs, has also been noted at Black Ven,[40] and the observation of such features supports the hypothesis of rotational failures. However, at Black Ven (photograph 127) the precise nature of much of the movement is masked by mudflows (photograph 129). The cliffs are a treacherous area of bog, muddy clay, water and vegetation through which there is a constant downward movement of mud and small stones. Superimposed on this ongoing denudation are a number of significantly lower frequency events, with Lang[41] describing a movement in 1958 when a quantity of Upper Greensand slipped down from the higher 'terraces' accompanied by a sound like thunder. There followed a sort of explosion in the cliff and a mass of Lias rocks were hurled out followed by a river of liquid clay which began to pour over the lower cliff to the shore. The event continued and produced a mudflow blocking the beach and ultimately produced a debris fan which formed a small promontory. Numerous fans of this kind at the seaward end of major gulley systems which have been incised into the surface of the mudflows are shown in photograph 128.

The situation at Cain's Folly, Stonebarrow (photograph 130) is less clear although rotational failures would seem to characterise movements of the upper cliff face. The middle cliff section is a disturbed area of mudslides which end abruptly in steep sea cliffs. A clearer picture of the relationship between the various types of mass movement is provided in the diagram on p. 254, where the detailed morphology of this landslide (known as the Fairy Dell) and the underlying lithology involved is illustrated.[42]

127 *opposite* Black Ven, Dorset. This general view shows the nature of the cliff scenery of west Dorset. Upper cliffs of Cretaceous rocks (Greensand) have slumped towards the sea on a mobile mass of Liassic marls, shales and clays. At Black Ven in particular, the undercliff is an area of highly disturbed strata and extensive mudflows. SY 357932, looking E.

128 Black Ven, Dorset. The terraced nature of the cliff, made up of marl and shale bands of the Lower Lias, capped with Upper Greensand of the Albian stage of the Lower Cretaceous, is clearly seen. The poor drainage of the marls and shales leads to perched water tables and high pore-water pressures, causing the seaward face of the cliff to slump. In places, significant mudflows descend across the lower cliffs forming arcuate promontories on the beach. On-going marine erosion gradually trims back the slumped masses, further undercutting the cliffs, leading to oversteepening and a continuation of the slumping process. SY 357932, looking NW.

129 *opposite* Black Ven, Dorset. This close-up of the cliff face shows in detail the scar where rotational sliding has occurred. Below, Liassic marls have slumped in a mix of small rotational slides and mudflows. The hummocky nature of the surface of the mudflows is clearly seen bottom left. SY 354932, looking NW.

Field surveys and air photograph analysis enabled an evolutionary model for the mass movements on the seaward face of Stonebarrow to be constructed.[43] A series of distinct episodic movements were identified with rapid and significant rotational failures of the upper cliff followed by block disruption. Minor failures also occur in this upper zone. From 1887–1964, a period which included the major catastrophic failure of 1942, the cliff top had receded at an average rate of 0.71 metres per year in the central part of the main scar. However, during the period 1946–69, the cliff top appears to have retreated at a rate of only 0.13 metres per year. The geomorphic mechanisms relating to these changing rates of denudation are linked to the erosional processes operating further downslope (Zone 2 in the diagram on p. 254) and also to the loading on tilted blocks by scree and talus material or further rotational landslide units.

The middle cliff area (Zone 2 in the diagram on p. 254) experiences a cyclical evolution where the retreat of the lower cliff, generated by wave action at the base, causes gullying and accelerated mudflow activity, maintaining a gradient across the platform below the middle cliff. Failures occur in the cliff as the mudslides remove material downslope, resulting in retreats of the order of 2 metres per year. The near vertical sea cliffs of the Lower Lias undergo parallel retreat at a rate of about 0.5 metres per year.

130 Cain's Folly, Stonebarrow, Dorset. Cain's Folly, also known as Fairy Dell, is a complex landslide made up of a number of inter-related morphodynamic zones. The scar of the upper cliff of Greensand is clearly seen in the middle of the photograph. Here, rotational sliding on Gault clay and Liassic marls towards the sea gives rise to disturbed and backward tilting strata. Downslope, the marls have slumped to form a middle zone of mudslides across a platform which ends with a steep drop to sea level and the sea-contact cliffs of the lower zone. Disturbed ground on the landward side of Stonebarrow, towards the Char Valley, can be seen centre right. These latter movements have now largely stabilised. SY 382928, looking W.

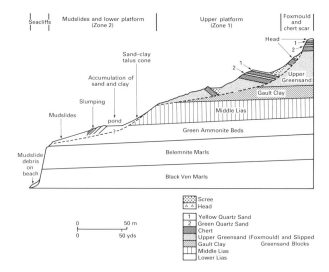

The Fairy Dell landslide complex at Stonebarrow Hill, Dorset.

254

Thus retreat at the base causes an increasing rate of debris transport across the lower platform undermining the middle cliff. Sea cliff retreat therefore activates a transgressive 'zone of aggression' which is transmitted inland.[44] The resultant major failures in the upper cliff can cause a reverse impulse for the settlement of new block failures; this controls and loads pre-existing landslide units which may be shunted downslope, thereby producing further failures in the middle cliff. Only when a considerable proportion of this material is removed across the lower platform is it possible for the cycle to begin again.

FOOTPATH EROSION AND DUNE ENCROACHMENT

Erosional and ecological changes that result from high levels of recreational activity have been causing concern to those interested in and responsible for the management of seminatural areas.[45] In this context, recreation can be defined as the pursuit of informal leisure in the countryside. The demand for recreation in the countryside continues to increase rapidly with a strong indication of a sharp rise in day trips.[46] The general effects of this increase are to damage the vegetation by trampling and to accelerate the loss of soil.

Most visitors seek landscapes with unusual features such as good views or specific features well known for their location, rock formation or classic landform type. Such popular 'beauty spots' range, for example, from Box Hill in Surrey on chalk downland, Cheddar Gorge and its limestone caves in the Mendips, the granite cliffs of Lands End in Cornwall, to the peaks of Snowdonia in Wales and the Cairngorms in Scotland. All such areas have their particular attraction and visitor pressure can come from both 'serious' hillwalkers, climbers and ramblers (who may or may not be back-packing) to casual Sunday afternoon visitors.

One of the most obvious forms of damage involving the loss of vegetation and soil leading to footpath erosion is illustrated in photographs 131, 132 and 133. These upland/montane sites are particularly vulnerable because vegetation in these areas has inherently slow rates of growth and recovery. Managers of these seminatural areas (in the cases illustrated: photograph 131, the Dartmoor National Park; 132, Snowdonia National Park; and 133 the National Trust) are concerned about such habitat damage and recognise a general problem that visitor pressure may ultimately lead to a reduction in amenity value and the level of recreational activity which the area can support. Although this might be seen as a positive attribute, in that visitors may go elsewhere, restoration of trampling damage may never be fully successful. Additionally, the problem may become worse at a new 'honeypot'.

The recognition of the erosional power of trampling has led to concepts of recreational carrying capacity, and there are two main aspects. Firstly, physical capacity, that is the maximum level of recreation use, in terms of the number of people or cars that can be accommodated, and secondly, ecological capacity, the maximum level of use that can take place before a decline in ecological value of the site or area. Although ecological values are difficult to assess, carrying capacity has been defined as the maximum intensity of use an area will continue to support under a particular management regime without inducing a permanent change in the biotic environment maintained by that management.[47]

At the local or site-specific scale, various attempts have been made to measure visitor numbers and their distribution. These techniques include questionnaire surveys, mechanical and optical counters.[48] The exact force exerted by trampling on any part of a path is not usually capable of measurement in the field except under tightly controlled experimental conditions. The weight of the walker, the type of footware, even the gait of the individual, all influence the detailed force distribution on the ground.[49] Particles on a path may be pushed, rolled or dragged along with the variation in trampling forces. Thus, simple counts of numbers of walkers on a path do not necessarily present a full picture and although path morphology is undoubtedly related to intensity of use, there may not be a direct linear relationship. The main factors involved in recreation pressure forces have been summarised as follows: (a) number of walkers, (b) slope of the ground modifying the effect of (a), and (c) distribution of walkers across a path.[50] Once a footpath has become established, a range of geomorphological factors become involved in the erosion process. These include: (i) the volume of run-off directed down a path and its source area, (ii) the infiltration capacity of the trampled soil on the path, (iii) the nature of the water flow down a path which may lead to gullying, (iv) slope and aspect of the path, which may be related to (v) exposure of the site to frost-heave and (vi) exposure of the site to wind effects.

131 *opposite* Haytor, Dartmoor, Devon. This is a classic Dartmoor summit tor. Its form is interesting displaying two distinct masses of granite with an avenue of more easily eroded material between the protruding granite bosses. This avenue and the surrounding clitter slope provide an easy incline up which thousands of visitors walk each year. Nearby access by road has made this site exceedingly popular as the numerous paths etched through the thin vegetation cover show. SX 755770, looking N.

132 Snowdon, Gwynedd, North Wales. The well-marked footpaths around Llyn Llydaw, which join with the Pyg Track at Glasslyn (centre of photograph) and the path up the arrête of Crib-goch, clearly show the main walkers' routes to the summit of the mountain. These paths are well-used during summer months and indicate the significant erosion that hill walkers can cause. SH 625545, looking SW.

133 Brimham Rocks, North Yorkshire. This area of Millstone Grit moorland on the eastern margin of the Yorkshire Dales is a popular tourist spot. The site is easily accessible and the rock outcrops are criss-crossed by numerous paths. Footpath erosion requires careful management at such popular sites. SE 210650, looking N.

Against these erosive forces can be listed the resistance factors of the soil involving considerations of particle size, stoniness and sheer strength. Detailed investigations indicate that the most actively eroding footpaths occur on higher slopes ($>17°$) and in *Calluna* dominated vegetation, and, 'It is in these sites that protective strategies are most urgently required, for example, by restricting lateral spread of walkers, or, in some cases, by re-routing paths away from sensitive areas.'[50]

The 'concentration' of hill walkers to well-defined paths on Snowdon is clearly demonstrated in photograph 132. The steepness of the terrain forces visitors to keep to an obvious path whether taking the Pyg Track route or the steeper climb up the arête of Crib-goch. In some senses, footpath erosion is less of a problem on Snowdon than on an upland site on Dartmoor such as Haytor (photograph 131). Bare rock surfaces, subjected to winter freeze-thaw weathering processes, are geomorphologically active and hence do not support extensive soil or vegetation cover, and whilst footpath erosion does occur on Snowdon, its scale could be described as insignificant compared with natural on-going geomorphological weathering and erosion.

Haytor, on the granite massif of Dartmoor, and Brimham Rocks on the Millstone Grit in North Yorkshire, demonstrate a more difficult situation for footpath erosion than that for the Snowdon area. Both of these sites have easy access by car and are visited throughout the year mainly by casual visitors. In a survey (1985) carried out by the Dartmoor National Park (pers. comm.) 60 per cent of visitors questioned were intending to walk on or around Haytor and 30 per cent described themselves as sightseers; nearly 40 per cent were only walking up to a mile. Thus a pattern emerges where visitors arrive by car, walk to the top of a hill or view point and return to their car. Sometimes, this activity may include a visit to a cafe or shop or even a picnic. This may seem apparently harmless but, as the network of footpaths on photograph 131 shows, the sheer number of visits is placing considerable pressure on such popular sites and there is a need for active management and possibly even some restriction. On Dartmoor, the National Park Authority has attempted to attract visitors to other sites or 'honeypots', and whilst this works to a certain extent there is still no real solution to visitor concentrations at well-known 'beauty-spots' or specific sites, as illustrated in the following table.

Number of visitors to Dartmoor: average Sunday and weekday
(one-way person trips crossing cordon, August 1985)

	SUNDAY	WEEKDAY
Recreation stop on Dartmoor		
Home based	9,200	5,000
Holiday-home based	8,100	9,700
Total recreation stops	17,300	14,700
Other purpose stops	4,500	6,000
Through trips	5,500	4,400
Total persons	27,300	25,100

Source: Dartmoor National Park, *1985 Car Park Interview Survey Report.*

An allied problem, where human activities are causing loss of vegetation cover, with consequent geomorphological erosion, is demonstrated by sand encroachment resulting from dune blow-outs in the area known as the Towans on the eastern side of St Ives Bay in West Cornwall. Here, the area between Hayle and Godrevy Point (photograph 134) has a long history of sand accumulation. Considerable quantities of quartz have been produced from the weathering and erosion of local granite rocks, most of which was delivered to the coastal zone

134 The Towans, near Hayle, Cornwall. Sand is blown onshore by westerly winds, accumulating throughout the post-glacial period. Parts of the dune complex have been stabilised by vegetation, although this is currently being damaged by holiday chalet development and tourist pressure. SW 555387, looking N.

during periods of climatic deterioration, such as the cold phases of the Pleistocene. During such cold phases all of St Ives Bay would have been a dry 'sand flat' across which inland rivers no doubt meandered, depositing the coarser sand-sized fraction of their load en route to the then distant 'off-shore' shoreline. There is also the possibility that an ice-front in the southern Irish Sea or even closer to the North Cornish Coast was providing glacial outwash sands.[52] Cemented sand rock below Devensian solifluction deposits at Godrevy Point indicates that, at least during the Ipswichian interglacial, sand was brought inland by a combination

of incoming wave action and wind transportation – a twin process which is being repeated in the modern post-glacial epoch.

Thus the sand dunes of the Towans have been built from sand blown onshore by predominantly westerly winds, and the dunes have subsequently become 'fixed' by a thin cover of soil and vegetation. However, with the absence of steeply rising ground inland (at least for some 2–3 km), natural blow-outs result in sand encroachment and subsequent dune migration across adjacent farmland.

Exacerbating the movement are two factors, firstly, at the northern end of the Towans, sand abstraction takes place. Here, the 'quarrying' activity of bull-dozers and the movement of vehicle traffic soon destroys any protective vegetation cover and the sand is free to blow inland. Secondly, the development of caravan and chalet sites, and now significant holiday accommodation complexes, has led to considerable modification of the dunes, as shown clearly in photograph 135. Many of the chalets have been grouped in the bottom of or adjacent to a blow-out. Vegetation has no chance to re-establish itself due to the concentration of both buildings and holiday makers. In addition, the construction of vehicle access further removes soil and vegetation. The fragile habitat and ecosystem of the dune is destroyed, opening the way for rapid and active sand movement.

135 Towans, north of Hayle, Cornwall. This is a closer view of the sand dune complex on the eastern side of St Ives Bay, where the thin nature of the soils and the sparse cover of vegetation is clear: blow-outs have also occurred. SW 560391, looking SE.

SOIL EROSION

Although the erosion of agricultural land in Britain has been observed from time to time, the general low rainfall erosivity regime experienced throughout the country has encouraged the view that water erosion on British arable soils is insignificant. One problem is that relatively few measurements of the rate of soil erosion have been published. Moreover, agricultural activities often remove direct evidence of soil erosion relatively quickly. Recent studies, however, have indicated that severe erosion has and does occur especially on bare arable soils which have been compacted ('capped') by rain drop impact.[53] It has also become apparent that low rainfall intensities may be offset by 'local' or 'site' factors such as soil texture and low organic matter content, slope length, steepness, type of cover and growth stage of crop, and removal of field boundaries. The effects on the soil, known collectively as degradation, reduce its value as a growing medium. Greater farm mechanisation with its notable increase in wheeled traffic has had a marked impact on the land; whilst the omission of grass leys from arable rotations has brought about a decline in soil organic content. Such changes in farming practice have extended the range of soil types which are prone to degradation. Soil degradation can also result in the loss of soil aggregate structure, enhancing the breakdown into primary particles and thus rendering the soil less resistant to erosion by water or wind.[54]

Soil erosion is identified as a two-stage process comprising firstly the detachment of soil particles from the ground mass and secondly their removal in a transporting medium. Such observations have led to the development of soil loss models and equations for erosion on sloping agricultural land. Initially, soil loss was related to slope steepness and length, but further developments led to the addition of a 'climatic' factor based on the maximum rainfall intensity over a 30-minute period. Subsequently, other factors have been incorporated to take account of the protection-effectiveness of different crops, conservation practices (e.g. contour ploughing), and an index of soil erosivity. These considerations have ultimately yielded the Universal Soil-Loss Equation.[55]

In recent models, a distinction is made between inter-rill erosion, which results from a combination of raindrop impact and overland flow and rill erosion where the runoff becomes channelled, the former being transport-limited with velocities around 0.001 m/s (for overland flow), whilst rill erosion, with velocities approaching 1 m/s, is detachment-limited (where the detachment is governed by soil particle size). In general, once a particle has been detached it can be transported until the velocity of the medium drops below a critical value. Such 'threshold' velocities vary with particle size, density and the cohesiveness of the soil.[56]

Soil erosion events are important in Britain in spite of the supposed existence of a stable landscape. The scale of these rill networks is quite small, and so too is the slope angle where they have developed. Although the area occupied by these examples is insignificant compared with the total land under cultivation, the frequency of such erosion occurrences is an important consideration in the assessment of the problem. Photograph 136 was taken during the early season period of crop cultivation and indicates the results of an ineffective vegetation cover offering little protection against early summer convective thunderstorms. Photograph 137, was also taken in early summer and indicates quite clearly the quantity of soil that can be moved after a significant precipitation event. Rainfall impacts have been affected by external influences such as the soils being too cold and moist for seed germination or soil compaction impeding crop growth and ploughing downslope has also exacerbated the situation.

Apart from the study of the mechanics of erosion in order to derive a soil erodibility index, research is now also concerned with both the growing period and the type of crop grown. Fullen and Harrison-Reed[57] point out that farm management systems which move the period when soils are bare towards summer periods, when convective storms and associated high intensity precipitation

136 Soil erosion at approximately 3 km SE of Yarmouth, Isle of Wight. This rill network has developed on sandy and silty soils to the east of the Yar valley, brought about by a combination of heavy rain and an ineffective vegetation cover. The problem has probably been made worse by the overworking of the soil and compaction of the surface by farm machinery. SZ 376860.

137 An erosive channel and rill system at Silsoe, Bedfordshire. This ground photograph, taken during the month of April, shows the extensive damage that can be brought about by fluvial soil erosion even on these gentle 5° slopes. The crop here is leeks, at an early growth stage, quite unable to offer protection from rainfall. Cultivation with the slope, although not the direct cause, has probably aided the development of this rill system. A factor in assessing the scale of such erosion on arable soils is that the evidence may be removed by subsequent reploughing. TL 078346, looking ESE.

occur, incur considerable risk of excessive levels of soil erosion. In addition, Morgan[58] indicates that growing of certain crops, especially large-leaved vegetables may enhance degradation and erosion rather than protect the soil, because the transformation of raindrops during interception, leading to impact on the soil by large drops, more than offsets the reduction in rainfall volume and energy provided by the plant cover. When this occurs surface crusting may result, enhancing overland flow or rill development. Thus a new generation of soil erosion models is required which take into account short-term or seasonal variation of soil strength and the nature of the eroding agent.

QUARRYING

Much larger scale impacts than footpath erosion, sand encroachment or soil erosion are shown in photographs 138 to 141, but even with quarry sites spoliation can be minimised. For example, dolerite (greenstone) abstraction from Carn Gwavas, between Newlyn and Mousehole in west Cornwall, is achieved with relatively little impact on the surrounding countryside. The rock is crushed on-site for aggregates and roadstone, and transported by coasters loaded at the adjacent jetty. The 'contained' nature of this site is indicated in photograph 138 and whilst landform modification has obviously taken place, viewed in the general context of a part of Britain well-known for vast tracts of former mining and quarrying sites, the impact of this quarry could be described as minimal.

On a slightly different scale is the well known 'hole' in North Cornwall, the site of the Delabole slate quarry (photograph 139). Excavated into a generally undulating planation surface, the quarry produces high quality slate much used as a speciality facing and roofing material throughout the south-west peninsula. However, as can be seen in the photograph, the search for slate of a suitable quality produces not only a deep excavation site, but more importantly some waste which is dumped adjacent to the site. This rock waste has encroached over the land around the quarry and represents the result of continuous extraction for a period of over 400 years. Currently, tipping of waste has almost ceased as more of the slate, particularly rock which does not split well, is marketed for patio or floor tiling, but the scars to the landscape remain.

Finally, perhaps one of the best examples of man-modified landscapes is provided by the extensive extraction of china clay in south-west England. Kaolinite, resulting from pneumatolysis of granite during late stages of cooling from the original molten rock material, is found principally in the Lands End Peninsula, at Tregonning Hill near Helston, on the St Austell Moors around Hensbarrow Beacon, on parts of Bodmin Moor and at Cornwood and Lee Moor on southern

138 Dolerite rock quarry at Carn Gwavas. This quarry on the south-western periphery of Newlyn is an important local source of roadstone. The impact of the site is quite small, viewed against the general rural hinterland of the Land's End peninsula. The rock is crushed and sorted on site, leaving little waste. SE 470279, looking W.

Dartmoor. The deposits have been worked for over 200 years and reserves are estimated to permit abstraction to continue for a further 100 years.

Some of the processes involved in the abstraction of the china clay are shown in photographs 140 and 141. High pressure jets break up the weakened or disaggregated granite and the kaolinite, mica and quartz sand slurry is washed to the bottom of the pit, sometimes at a depth of 100 metres. From here powerful pumps lift the slurry up inclined pipes to sand classifiers, mica pits and eventually to clay drying plants.

Clay production averages an 8:1 ratio of waste or other products to clay. Much of the waste is quartz sand which is tipped to form the familiar white conical hills which dominate the landscape. Despite attempts to use the sand as a building material or in the construction of 'concrete' blocks and other aggregates, English China Clays still have a problem with some 24 million tonnes of raw material – nearly all of which is destined to remain on or adjacent to the abstraction site. Local rivers run white with mica and fine sand waste, causing notable pollution problems downstream. Modification of the landscape is complete, with attempts to re-establish soils and vegetation on the waste tips. Hydraulic seeding mixed with fertilizers is reasonably successful and tree planting has also been undertaken. However, although these white hills may become green, they are still

139 *opposite* Delabole slate quarry, near Camelford, Cornwall. Delabole has been a centre for good quality slate production for over 400 years. As elsewhere in Britain, slate working produces quantities of waste material of poor quality for roofing: in the foreground new landforms are produced on the undulating planation surface by the construction of on-site waste tips. SX 075835, looking N.

140 China clay workings at Lee Moor, Devon. This site, on the southern fringe of Dartmoor, is the only large centre for the production of china clay in Devon and the impact is immediately apparent from the photograph. The method of working uses hydraulic jets to wash the kaolinised granite to the bottom of deep pits from which slurry is pumped to gravity separators. The process produces vast quantities of quartz gravel and sand, which despite its value to the construction industry is largely tipped on-site forming white conical hills. Such large-scale working is likely to continue for decades so that the environmental impact of the cones of waste sand will continue. SX 580623, looking S.

'unnatural' features of the landscape, and the extent and concentration of clay mining activity is such that continued production of waste, piled up into conical tips will be a feature of the changing landscape of the south-west peninsula for some time to come. In addition, whilst the steep slopes of the waste tips remain unvegetated, deep gullying occurs from active rill erosion of the unstable material on steep slopes. Even with full restoration and tree cover the tips remain as atypical landforms rising above the more gentle undulations of the upland surfaces of South West England.

GUIDE TO FURTHER READING

C. Embleton, D. Brunsden and D. K. C. Jones, *Geomorphology – Present Problems and Future Prospects*, Oxford, 1978.

A. Goudie, *The Nature of the Environment*, Oxford, 1984.

A. Goudie and R. Gardner, *Discovering Landscape in England and Wales*, London, 1985.

M. G. Hart, *Geomorphology – Pure and Applied*, London, 1986.

A. H. Perry, *Environmental Hazards in the British Isles*, London, 1981.

J. A. Steers, *Coastal Features of England and Wales*, Cambridge, 1981.

R. C. Ward, *Floods – A Geographical Perspective*, London, 1978.

J. B. Whittow, *Disasters – The Anatomy of Environmental Hazards*, Harmondsworth, 1980.

Notes

Chapter 1
Acknowledgement is made to Mr A. S. Burn (Senior Cartographer), Department of Geography, University of Southampton, for drawing maps and diagrams.

1 B. W. Sparks, *Rocks and Relief*, London, 1971.
2 A. F. Pitty, 'A study of some escarpment gaps in the southern Pennines', *Transactions of the Institute of British Geographers*, 36 (1965), 127–45.
3 J. B. Sissons, *Scotland*, London, 1976.
4 C. A. M. King, *Northern England*, London, 1976.
5 O. T. Jones, 'The drainage system of Wales and the adjacent regions', *Quarterly Journal of the Geological Society*, 107 (1951), 201–25.
6 E. H. Brown, *The Relief and Drainage of Wales*, Cardiff, 1960.
7 D. L. Linton, 'Geomorphology', in D. L. Linton (ed.), *Sheffield and its Region*, Sheffield, 1956, 24–43.
8 J. B. Sissons, 'The erosion surfaces and drainage system of south-west Yorkshire', *Proceedings of the Yorkshire Geological Society*, 29 (1954), 305–42.
9 Linton, 'Geomorphology'.
10 Sparks, *Rocks and Relief*.
11 Brown, *The Relief and Drainage of Wales*.
12 C. E. Everard, 'The Solent River: a geomorphological study', *Transactions of the Institute of British Geographers*, 20 (1954), 41–58.

Chapter 2
Acknowledgement is made to the Drawing Office Staff, Department of Geography, King's College, London, for preparing maps and diagrams.

1 T. F. Jamieson, 'On the parallel roads of Glen Roy, and their place in history of the Glacial period', *Quarterly Journal of the Geological Society of London*, 19 (1863), 235–59.
2 D. L. Linton, 'Watershed breaching by ice in Scotland', *Transactions of the Institute of British Geographers*, 15 (1949), 1–16.
3 G. S. Boulton, A. S. Jones, K. M. Clayton and M. J. Kenning, 'A British Ice Sheet Model and patterns of glacial erosion and deposition in Britain', in F. W. Shotton (ed.), *British Quaternary Studies*, Oxford, 1977.
4 K. M. Clayton, 'Zones of glacial erosion', in E. H. Brown and R. S. Waters (eds.), *Progress in Geomorphology*, Institute of British Geographers Special Publication 7 (1974), 163–76.
5 C. Embleton and C. Hamann, 'A comparison of cirque forms between the Austrian Alps and the Highlands of Britain', *Zeitschrift für Geomorphologie*, Supplementband 70 (1988), 75–93.
6 G. Manley, 'The late-glacial climate of north-west England', *Liverpool and Manchester Geological Journal*, 2 (1959), 188–215.
7 I. S. Evans, 'The geomorphology and morphometry of glacial and nival areas', in R. J. Chorley (ed.), *Water, Earth and Man*, London, 1969.

8 W. V. Lewis, 'The problem of cirque erosion', in W. V. Lewis (ed.), *Norwegian Cirque Glaciers*, Royal Geographical Society Research Series 4, 1960.
9 W. M. Davis, 'Glacial erosion in North Wales' *Quarterly Journal of the Geological Society of London*, 65 (1909), 281–350.
10 D. L. Linton, 'The forms of glacial erosion', *Transactions of the Institute of British Geographers*, 33 (1963), 1–28.
11 Linton, 'The forms of glacial erosion'.
12 A. M. D. Gemmell, 'The deglaciation of the Island of Arran, Scotland', *Transactions of the Institute of British Geographers*, 59 (1973), 25–39.
13 E. Watson, 'The glacial morphology of the Tal-y-llyn valley, Merionethshire', *Transactions of the Institute of British Geographers*, 30 (1962), 15–31.
14 D. E. Sugden, 'The selectivity of glacial erosion in the Cairngorm Mountains, Scotland', *Transactions of the Institute of British Geographers*, 45 (1968), 79–92.
15 D. E. Sugden, 'Landforms of deglaciation in the Cairngorm Mountains, Scotland', *Transactions of the Institute of British Geographers*, 51 (1970), 201–19.
16 A. Penck, 'Glacial features in the surface of the Alps', *Journal of Geology*, 13 (1905), 1–19, W. M. Davis, *Die erklärende Beschreibung der Landformen*, Leipzig, 1912.
17 Davis, 'Glacial erosion in North Wales'.
18 Linton, 'Watershed breaching by ice in Scotland'.
19 Linton, 'Watershed breaching by ice in Scotland'.
20 Penck, 'Glacial features in the surface of the Alps'.
21 J. Sölch, 'Studien über Gebirgspasse mit besonderer Berücksichtigung der Ostalpen: Versuch einer Klassifikation', *Forschungen zur deutschen Landes- und Volkskunde*, 17 (1908), 119–273.
22 Linton, 'Watershed breaching by ice in Scotland'.
23 S. Rudberg, 'Glacial erosion forms of medium size – a discussion based on four Swedish case studies', *Zeitschrift für Geomorphologie*, Supplementband, 17 (1973), 33–48.
24 Linton, 'The forms of glacial erosion'.
25 P. F. Kendall, 'A system of glacier lakes in the Cleveland Hills', *Quarterly Journal of the Geological Society of London*, 58 (1902), 471–571.

Acknowledgement is also made in respect of material for maps and diagrams drawn from the following sources:
p. 50, part from K. M. Clayton (1974); p. 64, E. Watson (1962); p. 66, D. L. Linton (1949).

Chapter 3
Acknowledgement is made to Mr G. Lewis (Senior Cartographer) and his assistants, Mr A. Lloyd and Mr P. Taylor, Department of Geography, University College of Swansea, for drawing maps and diagrams. Thanks are also expressed to Dr S. Campbell, Dr A. G. Dawson, Professor C. Embleton, Dr D. McCarroll and Dr J. E. Gordon for helpful comments on an early draft of this chapter and grateful thanks are made to Dr R. Cornish and Dr

D. F. M. McGregor for permission to refer to unpublished material.

1 J. D. Forbes, *Norway and its Glaciers Visited in 1851; and Excursions to the High Alps of Dauphine, Berne and Savoy*, Edinburgh, 1853.

2 A. C. G. Cameron, 'Note on a transported mass of chalk in the boulder clay at Catworth, Huntingdonshire, *Report, British Association for Advancement of Science*, 63 (1893), 760–1.

3 J. Geikie, *The Great Ice Age and its Relation to the Antiquity of Man*, 3rd edn, London, 1894.

4 Modified from E. H. Brown, 'The shape of Britain', *Transactions of the Institute of British Geographers*, new series, 4 no. 4 (1979), 449–62.

5 Cf. D. G. Sutherland, 'The raised shorelines and deglaciation of the Loch Long/Loch Fyne area, Western Scotland', unpubl. Ph.D. thesis, University of Edinburgh, 1981.

6 D. Q. Bowen *et al.*, 'Correlation of Quaternary glaciations in England, Ireland, Scotland and Wales', *Quaternary Science Reviews*, 5 (1986), 299–340. The term 'Wolstonian' is in quotation marks because there is currently (1988) a debate concerning its continued use. See, for example, J. Rose, 'Status of the Wolstonian Glaciation in the British Quaternary', *Quaternary Newsletter*, 53 (1987), 1–9; and P. L. Gibbard, and C. Turner, 'In defence of the Wolstonian stage', *Quaternary Newsletter*, 54 (1988), 9–14.

7 J. Lundqvist, 'Problems of the so-called Rogen moraine', *Sveriges Geologiska Undersökning*, C648 (1969), 1–32.

8 Cf. D. M. Hodgson, 'Hummocky and fluted moraines in part of North-West Scotland', unpubl. Ph.D. thesis, University of Edinburgh, 1982.

9 J. B. Sissons, Palaeoclimatic inferences from former glaciers in Scotland and the Lake District, *Nature*, 280 (1979), 199–202.

10 T. N. George, 'The glacial deposits of Gower, *Geological Magazine*, 70 (1933), 208–32.

11 N. Stephens, personal communication.

12 W. S. Symonds, *Records of the Rocks*, London, 1872.

13 T. M. Reade, 'The moraine of Llyn Cwm Llwch, on the Beacons of Brecon'. *Proceedings of the Liverpool Geological Society*, 7 (1894–5), 270–6.

14 C. A. Lewis, 'The periglacial landforms of the Brecon Beacons', unpubl. Ph.D. thesis, University College, Dublin, 1966.

15 J. B. Sissons, 'Palaeoclimatic inferences from former glaciers in Scotland and the Lake District', *Nature*, 280 (1979), 199–202.

16 N. Barlow, *The autobiography of Charles Darwin 1809–1882*, London, 1958.

17 A. G. Dawson, personal communication; K. Addison, *The Ice Age in Cwm Idwal*, 2nd edn, Broseley, Shropshire, 1988.

18 J. M. Gray, 'The last glaciers (Loch Lomond Advance) in Snowdonia, N. Wales', *Geological Journal*, 17 (1982), 111–33.

19 D. J. Unwin, 'Some aspects of the glacial geomorphology of Snowdonia', unpubl. M.Phil. thesis, University of London, 1970.

20 J. M. Gray, J. Ince and S. Lowe, 'Report on a short field meeting in North Wales', *Quaternary Newsletter*, 35 (1981), 40–4.

21 K. Addison, *Classic Glacial Landforms of Snowdonia, Landform Guide No. 3*, Sheffield, 1983.

22 J. B. Sissons, 'The Loch Lomond Advance in the Lake District, Northern England', *Transactions of the Royal Society of Edinburgh*, 66 (1980), 143–68.

23 Sissons, 'Palaeoclimatic inferences'.

24 Sissons, 'The Loch Lomond Advance'.

25 Bowen *et al.*, 'Correlation of Quaternary glaciations'.

26 G. S. Boulton and P. Worsley, 'Late Weichselian glaciation in the Cheshire–Shropshire Basin', *Nature*, 207 (1965), 704–6.

27 P. Worsley, 'Excursion report: the glacial geology of the Cheshire–Shropshire Plain', *Mercian Geologist*, 10 no. 3 (1986), 225–8.

28 F. W. Shotton (ed.), *The English Midlands: Guidebook for Excursion A2, INQUA Congress*, Norwich, 1977, 33–4.

29 E.g. M. Robinson and C. K. Ballantyne, 'Evidence for a glacial readvance pre-dating the Loch Lomond Advance in Wester Ross', *Scottish Journal of Geology*, 15 no. 4 (1979), 271–7.

30 C. K. Ballantyne, 'Protalus rampart development and the limits of former glaciers in the vicinity of Baosbheinn, Wester Ross', *Scottish Journal of Geology*, 22 no. 1 (1986), 13–25.

31 R. Cornish, 'Glacial geomorphology of the west-central Southern Uplands of Scotland, with particular reference to Rogen moraines', unpubl. Ph.D. thesis, University of Edinburgh, 1979.

32 G. S. P. Thomas, 'The Late Devensian glaciation along the border of northeast Wales', *Geological Journal*, 20 no. 4 (1985), 319–40.

33 D. F. M. McGregor, 'A quantitative analysis of some fluvioglacial deposits from east-central Scotland', unpubl. Ph.D. thesis, University of Edinburgh, 1974.

34 Cf. G. S. Boulton, 'Modern arctic glaciers as depositional models for former ice sheets', *Journal of the Geological Society of London*, 128 (1972), 361–93.

35 A. G. McLellan, 'The last glaciation and deglaciation of central Lanarkshire', *Scottish Journal of Geology*, 5 (1969), 248–68.

36 J. B. Sissons, 'The Perth Readvance in Central Scotland, part I', *Scottish Geographical Magazine*, 79 no. 3 (1963), 151–63, and 'The Perth Readvance in Central Scotland, part II', *Scottish Geographical Magazine*, 80 no. 1 (1964), 28–36.

37 P. F. Kendall, 'A system of glacier-lakes in the Cleveland hills', *Quarterly Journal of the Geological Society of London*, 58 (1902), 471–571.

38 J. B. Sissons, 'Some aspects of glacial drainage channels in Britain, part I', *Scottish Geographical Magazine*, 76 (1960), 131–46, and 'Some aspects of glacial drainage channels in Britain, part II', *Scottish Geographical Magazine*, 77 (1961), 15–36.

39 C. K. Ballantyne and D. G. Sutherland (eds), *Wester Ross; field guide*, Cambridge, 1987.

Acknowledgement is also made in respect of material for maps and diagrams drawn from the following sources:
p. 80, J. J. Lowe and M. J. C. Walker (1984); D. G. Sutherland and M. J. C. Walker (1984); C. K. Ballantyne and D. G. Sutherland (1987); J. Scourse (1987); p. 80, N. R. Eyles and W. R. Dearman (1981); p. 81, D. E. Sugden and B. S. John (1976); B. S. John (1979); p. 82, R. J. Price (1983); p. 82, R. J. Price (1973); p. 84, J. K. Charlesworth (1957); p. 90, J. M. Gray (1982); p. 88, K. Addison (1983); p. 90, J. B. Sissons (1980); p. 92, R. Cornish (1981); p. 94, M. A. Paul (1984); p. 100, M. Robinson and C. K. Ballantyne (1979); C. K. Ballantyne (1986); J. B. Sissons and A. G. Dawson (1981); p. 102, K. M. Clayton (1966); p. 104, G. S. P. Thomas (1985); p. 107, R. J. Price (1983); p. 112, J. B. Sissons (1982).

Chapter 4
Acknowledgement is made to Mr G. Lewis (Senior Cartographer) and his assistants, Mr A. Lloyd and Mr P. Taylor, Department of Geography, University College of Swansea, for drawing maps and diagrams; and for photographic work by Mr A. Cutliffe. Thanks is also expressed to Professor P. Worsley for providing the photographic print for photograph 58.

1 H. M. French, *The Periglacial Environment*, London, 1976.
2 A. L. Washburn, *Geocryology*, London, 1979.
3 J. A. Catt, 'Soil particle size distribution and mineralogy as indicators of pedogenic and gemorphological history: examples from the loessial soils of England and Wales', in K. S. Richards, R. R. Arnett and S. Ellis (eds), *Geomorphology and Soils*, London, 1985.
4 A. H. Lachenbruch, 'Mechanics of thermal contraction crack and ice-wedge polygons in permafrost', *Geological Survey of America Special Paper*, 70, (1962).
5 T. L. Péwé, 'Geomorphic processes in polar deserts', in T. L. Smiley and J. H. Zumberge (eds), *Polar Deserts and Modern Man*, Tucson, 1974.
6 L. King, 'High mountain permafrost in Scandinavia', *Proceedings of the Fourth International Permafrost Conference, Fairbanks*, Washington D.C., 1983.
7 C. K. Ballantyne, 'The present day periglaciation of upland Britain', in J. Boardman (ed.), *Periglacial Processes and Landforms in Great Britain and Ireland*, Cambridge, 1987.
8 J. R. Mackay, 'The world of underground ice', *Annals, Association of American Geographers*, 62 (1972), 1–22.
9 A. Lachenbruch, 'Contraction theory of ice wedge polygons: a qualitative discussion', *Proceedings of the First International Permafrost Conference*, National Academy of Science, National Research Council of Canada Publication 1287, 1966.
10 J. R. Mackay, 'Ice wedge cracks, Garry island, N.W.T.', *Canadian Journal of Earth Sciences*, 11 (1974), 1366–83.
11 R. F. Black, 'Les coins de glace et le gel permanent dans le Nord de l'Alaska', *Annales de Géographie*, 72 (1963), 257–71.
12 R. G. West, 'Stratigraphy of periglacial features in East Anglia and adjacent areas', in T. L. Péwé (ed.), *The Periglacial Environment*, Montreal, 1969. J. Rose, J. Boardman, R. A. Kemp and C. A. Whiteman, 'Palaeosols and the interpretation of the British Quaternary stratigraphy', in Richards *et al.*, *Geomorphology and Soils*.
13 F. W. Shotton, 'Large scale patterned ground in the valley of the Worcestershire Avon', *Geological Magazine*, 67 (1960), 404–8.
14 A. V. Morgan, 'Polygonal patterned ground of late Weichselian age in the area north and west of Wolverhampton, England', *Geografiska Annaler*, 53A (1971), 146–56.
15 A. M. D. Gemmel and I. B. M. Ralston, 'Some recent discoveries of ice-wedge cast networks in north-east Scotland', *Scottish Journal of Geology*, 20 (1983), 115–18.
16 R. B. G. Williams, 'Fossil patterned ground in eastern England', *Biuletyn Peryglacjalny*, 14 (1964), 337–49. A. S. Watt, R. M. S. Perrin and R. G. West, 'Patterned ground in Breckland: structure and composition', *Journal of Ecology*, 54 (1966), 239–58. F. H. Nicholson, 'Patterned ground formation and description as suggested by low Arctic and subarctic examples', *Arctic and Alpine Research*, 8 (1976), 329–42.
17 A. L. Washburn, 'Classification of patterned ground and review of suggested origins', *Geological Society of America Bulletin*, 67 (1956), 823–65.
18 Nicholson, 'Patterned ground formation'.
19 C. Tarnocai and S. C. Zoltai, 'Earth hummocks of the Canadian arctic and subarctic', *Arctic and Alpine Research*, 10 (1978), 581–94.
20 Nicholson, 'Patterned ground formation'. J. R. Mackay, 'The origin of hummocks, western Arctic coast, Canada', *Canadian Journal of Earth Sciences*, 17 (1980), 996–1006.
21 J. R. Mackay, 'Pingos of the Tuktoyaktuk Peninsula area, Northwest Territories', *Géographie Physique et Quaternaire*, 33 (1979), 3–61.
22 G. W. Holmes, D. M. Hopkins and H. L. Foster, 'Pingos in central Alaska', *U.S. Geological Survey Bulletin*, 1241A (1968).
23 Holmes *et al.*, 'Pingos in central Alaska'.
24 A. Pissart and P. Gangloff, 'Les palsas minerales et organique de la vallée de l'Aveneau, pres de Kuujjaq, Québec subarctique', *Géographie Physique et Quaternaire*, 38 (1984), 217–28.
25 B. W. Sparks, R. B. G. Williams and F. G. Bell, 'Presumed ground ice depressions in East Anglia', *Proceedings of the Royal Society of London*, 327A (1972), 329–43.
26 Sparks *et al.*, 'Presumed ground ice depressions in East Anglia'.
27 E. Watson, 'Investigations of some pingo basins near Aberystwyth, Wales', *Report 24th International Geological Congress (Montreal)*, Section 12 (1972), 212–33. E. Watson and S. Watson, 'Remains of pingos in the Cletwr Basin, southwest Wales', *Geografiska Annaler*, 56A (1974), 213–25.
28 Pissart and Gangloff, 'Les palsas minerales'.
29 C. Reid, 'On the origin of dry valleys and of coombe rock', *Quarterly Journal of the Geological Society of London*, 43 (1887), 364–73.
30 R. J. Small, 'The role of spring sapping in the formation of Chalk escarpment valleys', *Southampton Research Series in Geography*, 1 (1965), 1–29.
31 M. P. Kerney, E. H. Brown and T. J. Chandler, 'The late glacial and post glacial history of the Chalk escarpment near Brook, Kent', *Philosophical Transactions of the Royal Society of London*, 248B (1964), 135–204.
32 H. M. French, 'Asymmetrical slope development in the Chiltern Hills', *Biuletyn Peryglacjalny*, 21 (1972), 51–73.
33 C. Harris, *Periglacial Mass-Wasting: A Review of Research*, Norwich, 1981.
34 C. K. Ballantyne, 'The Late Devensian periglaciation of upland Scotland', *Quaternary Science Reviews*, 3 (1984), 311–43.
35 D. F. Ball and R. Goodier, 'Morphology and distribution of features resulting from frost-action in Snowdonia', *Field Studies*, 3 (1970), 193–218. Ballantyne, 'Present day periglaciation of upland Britain'.
36 J. P. Lautridou and J. C. Ozouf, 'Experimental frost shattering', *Progress in Physical Geography*, 6 (1982), 215–32.
37 'The Late Devensian periglaciation of upland Scotland'.
38 A. E. Corte, 'Particle sorting by repeated freezing and thawing', *Biuletyn Peryglacjalny*, 15 (1966), 175–240.
39 Corte, 'Particle sorting'.
40 R. J. Chandler, 'The inclination of talus, arctic talus terraces and other slopes composed of granular materials', *Journal of Geology*, 81 (1973), 1–14.
41 I. Statham, 'Scree slope development under conditions of surface particle movement', *Transactions of the Institute of British Geographers*, 59 (1973), 41–53.
42 Ballantyne, 'The Late Devensian periglaciation of upland Scotland'.
43 J. L. Innes, 'Lichenometric dating of debris-flow deposits in

the Scottish Highlands', *Earth Surface Processes and Landforms*, 8 (1983), 579–88.

44 J. M. Gray, 'The last glaciers (Loch Lomond Advance) in Snowdonia, N. Wales', *Geological Journal*, 17 (1982), 111–33. S. P. Oxford, 'Protalus ramparts, protalus rock glaciers and soliflucted till in the northwest part of the English Lake District', in J. Boardman (ed.), *Northern England, Field Guide to the Periglacial Landforms*, Cambridge, 1985, 38–46. C. K. Ballantyne and M. P. Kirkbride, 'The characteristics and significance of some Lateglacial protalus ramparts in upland Britain', *Earth Surface Processes and Landforms*, 11 (1986), 659–671.

45 J. B. Sissons, 'A remarkable protalus rampart in Wester Ross', *Scottish Geographical Magazine*, 92 (1976), 182–90.

46 C. E. Thorn, 'Quantitative evaluation of nivation in the Colorado Front Range', *Geological Society of America Bulletin*, 87 (1976), 1,169–78.

47 Harris, *Periglacial Mass-Wasting*.

48 P. J. Vincent and M. P. Lee, 'Snow patches on Farleton Fell, South-east Cumbria, *Geographical Journal*, 148 (1982), 337–42. C. K. Ballantyne, 'Nivation landforms and snow patch erosion on two massifs in the northern Highlands of Scotland', *Scottish Geographical Magazine*, 101 (1985), 40–9.

49 D. L. Linton, 'The problem of tors', *Geographical Journal*, 121 (1955), 470–87. J. Palmer and J. Radley, 'Gritstone tors of the English Pennines', *Zeitschrift für Geomorphologie*, 5 (1961), 37–52.

50 E. Derbyshire, 'Tors, rock weathering and climate in southern Victoria Land, Antarctica', in R. J. Price and D. E. Sugden (eds), *Polar Geomorphology*, Institute of British Geographers Special Publication 4, (1972), 93–105.

51 A. S. Goudie and N. R. Piggott, 'Quartzite tors, stone stripes, and slopes at the Stiperstones, Shropshire, England, *Biuletyn Peryglacjalny*, 28 (1981), 47–56.

52 D. F. Ball and R. Goodier, 'Large sorted stone-stripes in the Rhinog Mountains, North Wales', *Geografiska Annaler*, 50A (1968), 54–9.

53 Ballantyne, 'The Late Devensian periglaciation of upland Scotland'.

54 J. Warburton, 'Contemporary patterned ground (sorted stripes) in the Lake District', in Boardman (ed.), *Northern England, Field Guide to the Periglacial Landforms*. Ball and Goodier, 'Morphology and distribution'. Ballantyne, 'The Late Devensian periglaciation of upland Scotland'.

55 A. L. Washburn, 'Periglacial Problems', in M. Church and O. Slaymaker (eds), *Field and Theory, Lectures in Geocryology*, Vancouver, 1985.

56 R. J. Ray, W. B. Krantz, T. N. Caine and R. D. Gunn, 'A mathematical model for patterned ground: sorted polygons and stripes, and underwater polygons', *Proceedings of the Fourth International Permafrost Conference, Fairbanks*, Washington D.C., 1983.

Acknowledgement is also made in respect of material for maps and diagrams drawn from the following sources:
p. 116, T. L. Péwé (1974); p. 123, F. H. Nicholson (1976); p. 125, G. W. Holmes, D. M. Hopkins and H. L. Foster (1968); p. 126, B. W. Sparks, R. B. G. Williams and F. G. Bell (1972); E. Watson (1972); E. Watson and S. Watson (1974); p. 145, D. F. Ball and R. Goodier (1968); C. K. Ballantyne (1984); J. Warburton (1985); F. H. Nicholson (1976).

Chapter 5

Acknowledgement is made to Mr G. Lewis (Senior Cartographer) and his assistants, Mr A. Lloyd and Mr P. Taylor, Department of Geography, University College of Swansea, for drawing maps and diagrams; and for photographic work by Mr A. Cutliffe.

1 S. W. Wooldridge and R. S. Morgan, *The Physical Basis of Geography: An Outline of Geomorphology*, London, 1937.

2 A. N. Strahler, 'Quantitative analysis of watershed geomorphology', *Transactions of the American Geophysical Union*, 38 (1957), 913–20.

3 R. J. Chorley and B. A. Kennedy, *Physical geography: a systems approach*, London, 1971.

4 K. J. Gregory, 'Fluvial processes in British basins: the impact of hydrology and the prospect for hydrogeomorphology', in C. Embleton, D. Brunsden and D. K. C. Jones (eds), *Geomorphology: Present Problems and Future Prospects*, Oxford, 1978.

5 J. Lewin (ed.), *British Rivers*, London, 1981.

6 K. S. Richards, 'Evidence of Flandrian Valley alluviation in Staindale, North York Moors', *Earth Surface Processes and Landforms*, 6 (1981), 183–6.

7 D. K. C. Jones, *The Geomorphology of the British Isles: Southeast and Southern England*, London, 1981.

8 C. A. M. King, *The Geomorphology of the British Isles: Northern England*, London, 1976.

9 P. Cross and J. M. Hodgson, 'New evidence for the glacial diversion of the River Teme near Ludlow, Salop', *Proceedings of the Geological Association*, 86 no. 3 (1975), 313–31, and K. Hilton, *Process and Patterns in Physical Geography*, Slough, 1979.

10 King, *The Geomorphology of the British Isles: Northern England*.

11 K. S. Richards, *Rivers: Form and Process in Alluvial Channels*, London, 1982.

12 J. B. Sissons, 'The later lakes and associated fluvial terraces of Glen Roy, Glen Spean and vicinity', *Transactions of the Institute of British Geographers*, new series, 4 (1979), 12–29.

13 J. B. Sissons, *The Geomorphology of the British Isles: Scotland*, London, 1976.

14 Sissons, 'The later lakes and associated flurial terraces of Glen Roy, Glen Spean and vicinity'.

15 T. M. Thomas, 'Gully erosion in the Brecon Beacons area, South Wales, *Geography*, 41 (1956), 99–107.

16 R. P. D. Walsh, R. N. Hudson and K. A. Howells, 'Changes in the magnitude-frequency of flooding and heavy rainfalls in the Swansea Valley since 1875', *Cambria*, 9 no. 2 (1982), 36–60.

17 A. Jones, 'Rainfall, runoff and erosion in upper Tywi catchment', unpubl. Ph.D. thesis, University of Wales, 1975.

18 R. I. Ferguson, 'Channel form and channel changes', in J. Lewin (ed.) *British Rivers*, London, 1981.

19 Ferguson, 'Channel form and channel changes'.

20 J. Lewin and M. J. C. Weir, 'Morphology and recent history of the Lower Spey', *Scottish Geographical Magazine*, 93 (1977), 45–51.

21 A. Werrity and R. I. Ferguson, 'Pattern changes in a Scottish braided river over 1, 30 and 200 years', in R. A. Cullingford, D. A. Davidson and J. Lewin (eds), *Timescales in Geomorphology*, Chichester, 1981.

22 M. P. Mosley, 'Channel changes on the River Bollin, Cheshire, 1872–1973', *East Midland Geographer*, 6 (1975), 185–99.

23 K. J. Gregory, 'Drainage basin adjustments and Man', *Geographica Polonica*, 34 (1976), 155–73. C. C. Park, 'Man-induced changes in stream channel capacity', in K. J.

Gregory (ed.), *River Channel Changes*, Chichester, 1977.

24 G. E. Petts, 'Complex response of river channel morphology subsequent to reservoir construction', *Progress in Physical Geography*, 3 (1979), 329–62.

25 J. Lewin, 'Initiation of bedforms and meanders in coarse-grained sediment', *Bulletin of the Geological Society of America*, 87 (1976), 281–5.

Heavy metal concentrations in sediments in the lower Ystwyth floodplain at Llanilar and Wenallt[26]

RIVER REACH	TYPE OF LOCATION	METAL CONCENTRATIONS (mg/l)			
		Pb	Zn	Cu	Cd
Llanilar	Current channel sediments	274	480	27	1.3
	Current floodplain sediment	667	197	22	1.0
	Current floodplain sediment	384	225	20	0.9
	Old channel active post-1800	718	505	21	1.7
	Old channel active post-1800	6,017	730	129	1.7
	Ditto, but south of railway line	6,290	592	65	1.9
	Pre-1800 channel	443	483	25	1.5
Wenallt*	Current floodplain (381)	941	826	36	2.0
	Older floodplain (382)	2,321	707	80	1.9
	Older floodplain (383)	1,193	321	26	0.9
	Mill Race (384)	7,610	2,275	56	13.9
	Higher Terrace (379	670	226	22	0.9
	Valley slope (378)	71	111	20	0.7
	Valley slope (380)	65	145	17	0.7

*For location of samples see the diagram on p. 181.

26 J. Lewin, B. E. Davies and P. J. Wolfenden, 'Interactions between Channel Change and Historic Mining Sediments', in Gregory (ed.), *River Channel Changes*.

27 S. C. Bird, 'The effect of hydrological factors on past and present river pollution problems in South Wales, unpubl. Ph.D. thesis, University of Wales, 1986.

Changes in mean and annual maximum heavy metal concentrations (mg/l) in the Nant-y-Fendrod stream from 1965 to 1986

YEAR	ZINC		CADMIUM		LEAD	
	MAX.	MEAN	MAX.	MEAN	MAX.	MEAN
1965–66	38.4	11.9	0.94	0.10	2.3	1.17
1969–70	23.9	11.7	1.94	0.62	10.0	4.45
1974–75	12.3	9.1	0.16	0.09	0.3	0.15
1975–76	17.3	7.3	0.19	0.08	0.3	0.08
1979–80	74.0	9.0	0.42	0.08	1.4	0.10
1980–81	19.5	8.1	0.23	0.09	0.1	0.02
1981–82	16.8	5.8	0.23	0.06	0.1	0.04
1983–86	53.0	4.1	0.09	0.03	0.06	0.06
	Drinking water limits:					
Recommended	3.0		0.005		—	
Mandatory	0.5		0.001		0.05	

Sources: 1965–82 data from S.C. Bird
1983–86 data from Welsh Water Authority archive.

28 Bird, 'The effect of hydrological factors'.
29 P. G. Whitehead, S. C. Bird, R. J. Williams, C. Neal, R. Gray and R. Neale, *Strategies for Managing Nitrates and Heavy Metals in River Systems*, Institute of Hydrology, 1984.

30 M. D. Newson, 'Mountain streams', in Lewin (ed.), *British Rivers*.

31 J. H. Stoner, A. S. Gee and K. R. Wade, 'The effects of acidification on the ecology of streams in the upper Tywi catchment in West Wales', *Environmental Pollution, Series A*, 35 (1984), 125–57. J. H. Stoner and A. S. Gee, 'Effects of forestry on water quality and fish in Welsh rivers and lakes', *Journal of the Institute of Water Engineers and Scientists*, 39 (1985), 27–45.

Water quality of catchments of contrasting land use in the Llyn Brianne area of central Wales in September/October 1984.

	CATCHMENT		
WATER QUALITY PARAMETER	LI 1 Mature conifer	LI8 Young conifer	LI6 Moorland
Mean pH	4.87	5.25	6.96
Minimum pH	4.40	4.50	6.30
Mean aluminium (mg/l)	0.41	0.21	0.06
Maximum aluminium (mg/l)	1.31	0.72	0.43
Mean Total Hardness as CaCo$_3$ (mg/l)	6.3	7.7	15.6
Minimum Total Hardness as CaCo$_3$ (mg/l)	2.0	4.7	6.9
Mean Chloride (mg/l)	10.8	8.5	6.1
Minimum Chloride (mg/l)	9.0	8.0	5.0
Mean specific conductance (μS/cm)	59.0	50.0	55.0
Minimum specific conductance (μS/cm)	39.0	46.0	38.0

1. Sampling mostly daily: number of samples 39–41.
2. Locations of catchments: see the map on p. 184.

Acknowledgement is also made in respect of material for maps and diagrams drawn from the following sources:
p. 150, C. A. M. King (1976); p. 154, C. A. M. King (1976); p. 160, J. B. Sissons (1979); p. 163, T. M. Thomas (1956); p. 166, J. Lewin *et al.* (1977); p. 170, A. Werrity *et al.* (1981); p. 176, K. J. Gregory (1976), C. C. Park (1977); p. 181, J. Lewin *et al.* (in K. J. Gregory, 1977).

Chapter 6

Acknowledgement is made to Mr G. Lewis (Senior Cartographer) and his assistants, Mr A. Lloyd and Mr P. Taylor, Department of Geography, University College of Swansea, for drawing maps and diagrams; and for photographic work by Mr A. Cutliffe. Thanks is also expressed to Dr A. Mather and Dr R. A. Shakesby for helpful comments on an early draft of chapter 6.

1 J. A. Steers, *The Coastline of England and Wales*, Cambridge, 1946, and *The Coastline of Scotland*, Cambridge, 1973.

2 C. A. M. King, *Beaches and Coasts*, London, 1959; J. Pethick, *An Introduction to Coastal Geomorphology*, London, 1984; and E. C. F. Bird and M. L. Schwartz (eds), *The World's Coastline*, New York, 1985.

3 A. Robinson and R. Millward, *The Shell Book of the British Coast*, Newton Abbot, 1983, and J. L. Davies, *Geographical Variation in Coastal Development*, Edinburgh, 1972.

4 J. J. Lowe and M. J. C. Walker, *Reconstructing Quaternary Environments*, London, 1984; M. J. Tooley, *Sea-Level fluctuations in North-West England during the Flandrian Stage*, Oxford, 1978; and C. Kidson, 'The coast of south-west England', in C. Kidson and M. J. Tooley (eds), *The Quaternary History of the Irish Sea*, Liverpool, 1977.

5 R. J. N. Devoy, 'Flandrian sea-level changes in the Thames Estuary and the implications for land subsidence in England and Wales', *Nature*, 220 (1977), 712–15.

6 J. B. Sissons, *The Geomorphology of the British Isles: Scotland*, London, 1976.

7 J. M. Gray, 'Low-level shore platforms in the south-west Scottish Highlands: altitude, age and correlation', *Transactions of the Institute of British Geographers*, new series 3 (1978), 151–64.

8 A. G. Dawson, 'Shore erosion by frost: an example from the Scottish Late-glacial', in J. J. Lowe, J. M. Gray and J. E. Robinson (eds), *Studies in the Late-glacial of North-West Europe*, Oxford, 1980.

9 G. Jardine, 'The Quaternary marine record in south-western Scotland and the Scottish Hebrides', in Kidson and Tooley (eds), *The Quaternary History of the Irish Sea*.

10 G. F. Mitchell and A. R. Orme, 'The Pleistocene deposits of the Isles of Scilly', *Quarterly Journal of the Geological Society of London*, 123 (1969), 59–92.

11 J. D. Scourse (ed.), *The Isles of Scilly, Field Guide*, Quaternary Research Association, 1986.

12 Kidson, 'The coast of south-west England'.

13 Steers, *The Coastline of England and Wales*.

14 A. P. Carr and R. E. Baker, 'Orford, Suffolk: evidence for the evolution of the area during the Quaternary', *Transactions of the Institute of British Geographers*, 45 (1968), 107–23.

15 *Orford Ness: A Selection of Maps Mainly by John Norden*, Cambridge, 1966.

16 Steers, *The Coastline of Scotland*, and J. B. Whittow, *Geology and Scenery in Scotland*, Harmondsworth, 1977.

17 A. S. Mather and W. Ritchie, *The Beaches of the Highlands and Islands of Scotland*, The Countryside Commission for Scotland, 1977, and W. Ritchie, *The Beaches of Barra and the Uists*, The Countryside Commission for Scotland, 1971, 63–68.

18 S. N. Haynes and M. G. Coulson, 'The decline of *Spartina* in Langstone Harbour, Hampshire', *Proceedings of the Hants Field Club and Archaeological Society*, 38 (1982), 5–18.

19 M. A. Arber, 'Cliff profiles in Devon and Cornwall', *Geographical Journal*, 114 (1949), 191–97.

20 A. S. Goudie and R. Gardner, *Discovering Landscape in England and Wales*, London, 1985, and D. Dalzell and E. M. Durrance, 'The evolution of the Valley of the Rocks', *Transactions of the Institute of British Geographers*, new series, 5 no. 1 (1980), 66–79.

21 D. J. Lees, 'The evolution of Gower Coasts', in G. Humphrys (ed.), *Geographical Excursions from Swansea, Vol. I, Physical Environment*, 1982.

22 Goudie and Gardner, *Discovering Landscape in England and Wales*.

23 J. N. Hutchinson, 'Various forms of cliff instability arising from coast erosion in south-east England', *Geoteknikkdagen*, 19 (1979), 1–32, and C. L. So, 'Coastal Platforms of the Isle of Thanet, Kent', *Transactions of the Institute of British Geographers*, 37 (1965), 147–56.

24 J. N. Hutchinson, 'Field and laboratory studies of a fall in Upper Chalk cliffs at Joss Bay, Isle of Thanet', *Stress–Strain Behaviour of Soils*, London, 1972.

25 Goudie and Gardner, *Discovering Landscape in England and Wales*, and J. N. Hutchinson, E. N. Bromhead and J. F. Lupini, 'Additional observations on the Folkestone Warren landslides', *Quaterly Journal of Engineering Geology, London*, 13 (1980), 1–31.

26 J. N. Hutchinson, 'A pattern in the incidence of major coastal mudslides', *Earth Surface Processes*, 8 (1983), 391–97.

27 J. A. Steers, *The Sea Coast*, London, 1954, and Whittow, *Geology and Scenery in Scotland*.

28 A. W. Pringle, 'Beach development and coastal erosion in Holderness, North Humberside', in J. Neale and J. Flenley (eds), *The Quaternary in Britain*, 1981.

Acknowledgement is also made in respect of material for maps and diagrams drawn from the following sources:
p. 195, J. B. Sissons (1976); p. 200, G. F. Mitchell and A. R. Orme (1969); p. 206, J. A. Steers (1946); p. 209, A. P. Carr and R. E. Baker (1968); p. 212, J. A. Steers (1973); J. B. Whittow (1977); p. 213, A. S. Mather and W. Ritchie (1977); W. Ritchie (1971); p. 215, S. N. Haynes and M. G. Coulson (1982); p. 220, M. A. Arber (1949); A. S. Goudie and R. Gardner (1985); p. 223, D. J. Lees (1982); p. 225, A. S. Goudie and R. Gardner (1985); p. 226, J. N. Hutchinson (1979); C. L. So (1965); p. 227, J. A. Steers (1954); J. B. Whittow (1977); p. 229, J. A. Steers (1946); p. 231, A. W. Pringle (1981).

Chapter 7
Acknowledgement is made to the Staff of the Cartographic Resources Unit of the Department of Geographical Sciences, Plymouth Polytechnic for drawing the maps and diagrams.

1 W. M. Davies, 'The geographical cycle', *Geographical Journal*, 14 (1899), 481–504.

2 R. J. Chorley, 'The drainage basin as a fundamental geomorphic unit', in R. J. Chorley (ed.), *Water Earth and Man*, London, 1969.

3 K. J. Gregory, 'A physical geography equation', *National Geographer*, 12 (1978), 137–41.

4 K. J. Gregory, *The Nature of Physical Geography*, London, 1985.

5 R. C. Ward, *Floods – A Geographical Perspective*, London, 1978.

6 Natural Environment Research Council (NERC), *Flood Studies Report*, (5 volumes), London, 1975.

7 D. W. Reed, *A Review of British Flood Forecasting Practice*, Wallingford, 1984.

8 M. C. Jackson, 'The influence of snowmelt on flood flows in rivers', *Journal of the Institute of Water Engineers and Scientists*, 32 (1978), 495–508.

9 Jackson, 'The influence of snowmelt'.

10 NERC, *Flood Studies Report*.

11 Jackson, 'The influence of snowmelt'.

12 Reed, *A Review of British Flood Forecasting Practice*.

13 A. S. Goudie, *The Nature of the Environment*, Oxford, 1984.

14 N. A. Mörner, 'Eustacy and geoid changes', *Journal of Geology*, 84 (1976), 123–51.

15 See chapter 6, this volume.

16 R. W. Fairbridge, 'The changing level of the sea', *Scientific American*, 202 (1960), 70–9, and 'Eustatic changes in sea level', *Physical Chemistry of the Earth*, 4 (1961), 99–185.

17 S. Jelgersma, 'Holocene sea level changes in the Netherlands', *Meded geol. Sticht.*, (ser C6) 7 (1961), 1–100.

18 A. B. Hawkins, 'The late Weichselian and Flandrian Transgression of South West Britain', *Quaternaria*, 14 (1971), 115–30; Hawkins, 'Depositional characteristics of estuarine

alluvium: some engineering implications', *Quarterly Journal of Engineering Geology*, 17 (1984), 219–34; and A. Heyworth and C. Kidson, 'Sea level changes in southwest England and Wales', *Proceedings of the Geologists' Association*, 93 (1982), 91–111.

19 C. Thomas, *Exploration of a Drowned Landscape: Archaeology and History of the Isle of Scilly*, London, 1985.

20 H. Valentin, 'Present vertical movements of the British Isles', *Geographical Journal*, 119 (1953), 299–305.

21 G. H. Willcox, 'Problems and possible conclusions related to the history and archaeology of the Thames in the London Region', *Transactions of the London and Middlesex Archaeological Society*, 126 (1975), 285–92.

22 R. W. Horner, 'The Thames tidal flood risk – the need for the barrier: a review of its design and construction', *Quarterly Journal of Engineering Geology*, 17 (1984), 199–206.

23 K. F. Bowden, 'Storm surges in the North Sea', *Weather*, 8 (1953), 82–4.

24 J. B. Wittow, *Disasters – The Anatomy of Environmental Hazards*, Harmondsworth, 1980.

25 J. R. Rossiter, 'The great tidal surge of 1953', *The Listener*, 8 July, 1954, 55–6.

26 Rossiter, 'The great tidal surge of 1953'.

27 J. A. Steers, 'The east coast floods January 31–February 1 1953', *Geographical Journal*, 119 (1953), 280–98.

28 H. A. P. Jensen, 'Tidal inundations past and present', *Weather*, 8 (1953), 108–13.

29 G. Reynolds, 'Storm surge research', *Weather*, 8 (1953), 101–8, and A. J. Bowen, 'The tidal regime of the River Thames; long term trends and possible causes'. *Philosophical Transactions of the Royal Society, London*, (A) 272 (1972), 187–99.

30 D. K. C. Jones, *Southeast and Southern England*, London, 1981.

31 A. S. Goudie and R. Gardner, *Discovering Landscape in England and Wales*, London, 1985.

32 M. A. Arber, 'Landslips near Lyme Regis', *Proceedings of the Geologists' Association*, 84 (1973), 121–35, and W. A. Macfadyen, *Geological Highlights of the West Country*, London, 1970.

33 J. Pitts and D. Brunsden, 'A reconsideration of the Bindon landslide of 1839', *Proceedings of the Geologists' Association*, 98 (1987), 1–19.

34 Arber, 'Landslips near Lyme Regis'.

35 J. Pitts, 'An historical survey of the landslips of the Axmouth – Lyme Regis undercliffs, Devon', *Proceedings of the Dorset Natural History and Archaeology Society*, 103 (1981), 101–6.

36 Pitts and Brunsden, 'A reconsideration of the Bindon landslide of 1839'.

37 M. A. Arber, 'The coastal landslips of south-east Devon', *Proceedings of the Geologists' Association*, 51 (1940), 257–71.

38 Arber, 'Landslips near Lyme Regis', and 'The coastal landslips of south-east Devon'.

39 Arber, 'The coastal landslips of south-east Devon'.

40 Arber, 'The coastal landslips of south-east Devon'.

41 W. D. Lang, 'Report on Dorset Natural History for 1958: Geology', *Proceedings of the Dorset Natural History and Archaeology Society*, 80 (1959), 22.

42 D. Brunsden and D. K. C. Jones, 'The evolution of landslide slopes in Dorset', *Philosophical Transactions of the Royal Society, London*, (A) 283 (1976), 605–631, and Arber, 'The coastal landslips of West Dorset', Proceedings of the Geologists' Association 52 (1941), 273–283.

43 Brunsden and Jones, 'The evolution of landslide slopes in Dorset'.

44 Brunsden and Jones, 'The evolution of landslide slopes in Dorset'.

45 F. B. Goldsmith, 'Ecological effects of visitors in the countryside', in F. B. Goldsmith and A. Warren (eds), *Conservation in Practice*, London, 1974, and Goldsmith, 'Ecological effects of visitors and the restoration of damaged areas', in F. B. Goldsmith and A. Warren (eds), *Conservation in Perspective*, London, 1983.

46 M. Dower, 'Fourth Wave: the challenge of leisure', *Civic Trust*, 5 (1965); M. Dower, 'Amenity and tourism in the countryside', in J. Ashton and W. H. Long (eds), *The Remoter Rural Areas of Britain*, Edinburgh, 1972; and J. A. Patmore, *Land and Leisure*, Newton Abbot, 1970.

47 Goldsmith, 'Ecological effects of visitors in the countryside.

48 Dower, 'Amenity and tourism'; T. F. Wood, 'The analysis of environmental impacts resulting from summer recreation in the Cairngorm ski area, Scotland', *Journal of Environmental Management*, 25 (1987), 271–84; N. G. Bayfield, 'A simple method for detecting variations, in walker pressure laterally across paths', *Journal of Applied Ecology*, 8 (1971), 533–5; A. M. Coker and P. D. Coker, 'Some practical details of the use of pressure sensitive counters', *Recreation News Supplement*, 7 (1972), 14–17; and A. M. Coker and P. D. Coker, 'A simple method of time-lapse photography for use in recreation studies', *Recreation News Supplement*, 8 (1973), 31–8.

49 F. B. Goldsmith, R. J. C. Munton and A. Warren, 'The impact of recreation on the ecology and amenity of semi-natural areas: methods of investigation used in the Isles of Scilly', *Biological Journal of the Linnean Society*, 2 (1970), 287–306, and R. Coleman, 'Footpath erosion in the English Lake District', *Applied Geography*, 1 (1981), 121–31.

50 Coleman, 'Footpath erosion in the English Lake District'.

51 Coleman, 'Footpath erosion in the English Lake District'.

52 G. F. Mitchell, 'The Pleistocene history of the Irish Sea', *Advancement of Science*, 17 (1960), 313–25, and N. Stephens, 'The West Country and Southern Ireland', in C. A. Lewis (ed.), *The Glaciations of Wales and Adjoining Regions*, London, 1970.

53 M. A. Fullen and A. Harrison-Reed, 'Rainfall, runoff and erosion on bare arable soils in East Shropshire, England', *Earth Surface Processes and Landforms*, 11 (1986), 413–25.

54 J. Boardman, 'Soil erosion at Albourne, West Sussex, England', *Applied Geography*, 3 (1983), 317–29.

55 W. H. Wischmeier and D. D. Smith, 'Soil loss estimation as a tool in soil and water management planning', *International Association of Scientific Hydrology*, 59 (1962), 148–59.

56 F. Hjulstrom, 'Studies of the morphological activity of rivers as illustrated by the River Fyris', *Bulletin of the Geological Institute of the University of Uppsala*, 25 (1935), 221–527, and J. Savat, 'Common and uncommon selectivity in the process of fluid transportation: field observations and laboratory experiments on bare surfaces', in D. H. Yaalon (ed.), *Aridic Soils and Geomorphic Processes*, Catena Suppl. No. 1 (1982), 139–60.

57 Fullen and Harrison-Reed, 'Rainfall, runoff and erosion'.

58 R. P. C. Morgan, 'Soil degradation and erosion as a result of agricultural practice', in K. S. Richards, R. R. Arnett and S. Ellis (eds), *Geomorphology and Soils*, London, 1985.

Acknowledgement is also made in respect of material for maps and diagrams drawn from the following sources:
p. 239, H. Valentin (1953); p. 239, R. W. Horner (1984); p. 240, K. F. Bowden (1953); p. 246, R. W. Horner (1953); p. 247, M. A. Arber (1973); p. 247, M. A. Arber (1941, 1973); p. 251, W. A. Macfadyen (1970); p. 254, D. Brunsden *et al.* (1976).

Notes on contributors

Nicholas Stephens, formerly Professor of Geography at the University College of Swansea, graduated from the University of Bristol and completed his PhD at The Queen's University of Belfast. His main teaching and research interests have been in Quaternary geomorphology, mass-movement phenomena and environmental hazards. He has been President of Section E (Geography) of The British Association for the Advancement of Science, and taught in Belfast and the University of Aberdeen, and was Head of Department at Swansea, 1981–8.

Clifford Embleton is Professor of Geography at King's College London. He graduated at St John's College Cambridge and completed his PhD there, before taking up an appointment at the University of London. His main teaching and research interests have been in glacial and periglacial geomorphology. He is President of the Study Group on Geomorphological Hazards of the International Geographical Union and has been Visiting Professor in Austria, Canada and the USA.

Charles Harris is a Senior Lecturer in Geology at the University of Wales, Cardiff. He graduated from the University of Durham and completed his PhD at the University of Reading. His teaching career began in the Geography Department at University College of Swansea before transferring to the Geology Department at Cardiff. His main teaching and research interests are in glacial and periglacial geomorphology and Quaternary geology, and he has conducted geomorphological studies on landforms and sediments in Britain, Norway and Colorado, USA.

Richard Shakesby is a Lecturer in Geography at the University College of Swansea. He graduated from Portsmouth Polytechnic and completed his PhD at the University of Edinburgh in glacial geomorphology. He has published research on glacial and periglacial landforms and sediments in Britain and Norway, and is currently engaged in geomorphological studies in Portugal and Zimbabwe concerned with aspects of Late Quaternary and present day rates of soil erosion.

Peter Sims is Principal Lecturer in Physical Geography at Plymouth Polytechnic. He graduated from Portsmouth Polytechnic and obtained an MSc in hydrogeology at the University of London. Since 1969 he has taught at Plymouth Polytechnic, where his main teaching and research have been in applied topics, including hydrology, water resources and coastal zone management. He has also been involved in geomorphological studies, specialising in Quaternary landform evolution in south west England.

John Small is Professor of Geography at the University of Southampton. He graduated and took his PhD at Queens' College Cambridge and has taught at the University of Southampton for the whole of his career. He has completed major research in the geomorphology of the English chalklands and in recent years has studied active processes of erosion, transport and deposition by Alpine glaciers, mainly in Switzerland.

Rory Walsh is a Lecturer in Geography at the University College of Swansea. He graduated and took his PhD at St John's College Cambridge and has been a lecturer at Swansea since 1976, where his main teaching and research interests are in tropical geomorphology and hydrology. He has worked in the West Indies, the Sudan and Sarawak and his other interests include river pollution, acid rainfall and runoff, river flooding, and soil erosion in Portugal.

Photographs

All the aerial photographs in this volume are in the Cambridge University Collection of Aerial Photography and are used with kind permission. They are in the copyright of the University of Cambridge, with the exception of those noted below. The C.U.C.A.P. photo reference is given in brackets.

1 Suilven, Sutherland (BVJ 68)
2 The Malvern Hills, Herefordshire (CAW 96)
3 Rhinog Fawr, Gwynedd, North Wales (ASR 70)
4 Wenlock Edge, Shropshire (CEQ 79)
5 Glen Mor, Inverness-shire (BRD 90)
6 Borrowdale, West Cumberland (CCL 93)
7 Brecon Beacons (BLH 31)
8 (a) Giggleswick Scar, Nr Settle, Yorkshire (BHG 38)
8 (b) Malham Cove, Nr Settle, Yorkshire (EM 031) Crown Copyright
9 Curbar Edge, Derbyshire (BWH 019)
10 Haytor Rocks, Dartmoor, Devon (BFF 58)
11 Skomer Island, Dyfed, West Wales (BLI 37)
12 The Whin Sill, Nr Haltwhistle, Northumberland (ATN 14)
13 Purbeck Ridge, Dorset (UL 38) Crown Copyright
14 Hambleton Hills, Yorkshire (AZZ 99)
15 The South Downs at Ditchling Beacon, East Sussex (BOS 30)
16 Western Mull, Scotland (QX 64) Crown Copyright
17 The New Forest, Hampshire (AAS 46)
18 Cwm Cau and Cader Idris, North Wales (K17-P 262)
19 Cwm Cau and Cader Idris, North Wales (PZ 53/54) Crown Copyright
20 Mynydd Mawr, Snowdon, North Wales (EJ 035) Crown Copyright
21 Snowdon, North Wales (EJ 049) Crown Copyright
22 Lliwedd, near Snowdon, North Wales (EJ 047) Crown Copyright
23 Coir-a-Ghrunnda, Cuillin Hills, Skye (BHA 59)
24 Loch Coruisk, Skye (BHA 54)
25 Glen Rosa, Arran, Scotland (WD 31) Crown Copyright
26 Tal-y-llyn Valley, West Wales (EH 040) Crown Copyright
27 An Garbh Coire, Cairngorm Mountains (BUD 099/100)
28 Tail Burn and Moffat Water, S. Scotland (BGZ 55)
29 Nantlle Valley, North Wales (CEO 55)
30 Lairig Ghru, Cairngorm Mountains, Scotland (RD 68) Crown Copyright
31 Loch Avon, Cairngorm Mountains, Scotland (AHZ 36)
32 Loch Laxford, Sutherland (CEL 12)
33 Aberglaslyn Pass, North Wales (EJ 056) Crown Copyright
34 Newtondale, East Yorkshire (BA 30)
35 Ailsa Craig, Firth of Clyde (BOO 53)
36 Cwm Llwch, Brecon Beacons, South Wales (BIE 18)
37 Cwm Crew, Brecon Beacons, South Wales (BLH 34)
38 Cwm Idwal, Snowdonia, North Wales (ABD 67)
39 Bowscale Fell, Lake District (BVY 076)
40 Mullwharchar, Ayrshire (UW 18) Crown Copyright
41 Elgin Lane, Kirkcudbrightshire/Ayrshire (UW 17) Crown Copyright
42 Upper Oxendale, Lake District, Cumbria (RC8-BO 166)
43 Ellesmere–Whitchurch moraine, Shropshire (BNK 41)
44 Applecross moraine, Wester Ross, north-west Scotland (BSQ 034)
45 Baosbheinn ridge, Wester Ross, north-west Scotland (BSQ 021)
46 Drumlins, Ribblehead, Yorkshire (AWK 20)
47 Rogen moraine, Carsphairn Valley, south-west Scotland (MR 091) Crown Copyright
48 Kettle holes, NE of Wrexham, Clwyd (CKS 94)
49 Kame complex, NE of Cairn Water, Dumfriesshire (CGA 046)
50 The Bedshiel esker, Greenlaw Moor, Berwickshire (CAR 83)
51 Carstairs esker–kame complex, Lanarkshire (BKW 59)
52 West Water, Angus (ATP 65)
53 Camelon, east-central Scotland (BQN 60)
54 Achnasheen, central Ross-shire (RC8-CG 92/98/100)
55 Achnasheen, central Ross-shire
56 Top of an ice wedge exposed by placer mining in river gravels, Hunker Creek, Klondike District, Yukon Territory, Canada
57 Low-centre ice-wedge polygons, Mackenzie Delta, North West Territories, Canada
58 Two ice-wedge casts, Baston, Lincolnshire
59 Fossil ice-wedge polygons, near Adwick-Le-Street, Yorkshire (AZC 74)
60 Large scale non-sorted chalkland patterned ground, near Elveden, Suffolk (AKU 80)
61 Large scale non-sorted chalkland patterned ground, Thetford Heath, Suffolk (AGX 17)
62 Small closed-system pingo, Mackenzie Delta North-West Territories, Canada
63 Fossil ground ice depressions, possibly open-system pingos, Walton Common, Norfolk (AWD 73)
64 Fossil ground ice depressions, SW of Llanpumpsaint, Carmarthenshire (CET 61)
65 Dry valley in chalk escarpment, Barton in the Clay, Bedfordshire/Hertfordshire border (AP 67)
66 Dry valley in chalk, SE of Amberley, Sussex (AIS 46)
67 Panorama of the Cheviot Hills, near Rochester, Northumberland (BAJ 1)
68 Oxwich Point, Gower, South Wales (BXE 62)
69 Afon Sychlwch, the Old Red Sandstone escarpment of the Brecon Beacons, South Wales (CKX 31)
70 Lairig Ghru, Cairngorm Mountains, Scotland (BOW 42)
71 Sròn na Lairige, Braeriach, Cairngorm Mountains, Scotland (BOW 21)
72 Beinn Eighe, Ross and Cromarty, Scotland (BSP 077)
73 An Garbh Choire, Cairngorms, Scotland (BUD 097)

Glossary

aggradation The building-up of the land surface by the deposition and accumulation of sediments, derived from either river erosion or the deposition of marine sediments.

albedo The reflection coefficient or reflectivity of an object; the ratio between total solar electro-magnetic radiation (short wavelengths) falling upon a surface and the amount reflected, expressed as a decimal or percentage. Average albedo of the Earth is about 0.34 to 0.4 (34 to 40 per cent) but it varies according to colour and texture of the surface.

Allerød A late-glacial interstadial phase, dating from approximately 12,000 to 10,800 BP (also known as Pollen Zone II).

alluvial fan A fan-shaped or cone-shaped deposit of sediments laid down by a river, especially where it emerges from a steep constricting valley on to a flatter plain.

Anglian A glacial stage of the Middle Pleistocene represented by deposits at Corton, Suffolk and believed to pre-date the 'Hoxnian' interglacial and 'Wolstonian' glacial periods.

antecedent precipitation The amount of precipitation that has fallen in the previous seven, five or three days, sometimes as an index value.

anticlinal An arched fold or upfold in rock strata (anticline) where the two sides (limbs) of the fold dip in opposite directions away from the crestline of the central axis.

arcuate Possessing an arc-like form in plan view, such as a delta, spit, etc.

arenaceous rocks A group of detrital sedimentary rocks, typically sandstones, in which the predominant minerals are quartz, feldspars, mica, glauconite and iron oxides.

arête A steep-sided rocky ridge, especially in high mountains, where it may form the crest between two adjacent cirques (German: *Grat*).

argillaceous A rock composed largely of clay minerals, the resultant rock types including clays, mudstone, marl, shale and slate.

Atlantic A climatic period of the Holocene dating from about 7,500 to 5,200 BP; known as the 'climatic optimum' and characterised by temperatures 2° to 3° above those of the present. There was extensive growth of mixed oak forest, widespread peat formation and a rise in sea level of a few metres (Pollen Zones VI-VII in Britain).

BP Before the present (taken as AD 1950). A term widely used in the radiometric dating of glacial and post-glacial time in preference to BC. See radiocarbon dating.

batholith A large mass of intrusive igneous rock (usually granite), formed by the deep-seated emplacement of magma. The domed upper surface may be later exposed by denudation to form substantial uplands, e.g. Dartmoor.

bedload The solid material carried along the bed of a river by saltation and gravity; particles are pushed and rolled along the stream bed.

blow-out A hollow in sand dunes, especially along the coast, where wind eddying has enlarged a break in the protective vegetation.

Boreal A climatic period of the Holocene dating from about 9,500 to 7,500 BP. Generally a dry phase with cold winters and warm summers, indicated by (in Britain) a birch–pine–hazel flora; period spans Pollen Zones V and part of VI in Britain.

braiding Refers to a river course consisting of a network of inter-connected converging and diverging channels, with banks of alluvial material between them.

breccia A rock composed of angular rock fragments (clasts) usually cemented together in a matrix.

brown earth One of the main zonal soil groups, characteristic of areas in middle latitudes; formerly covered with deciduous woodland, rich in organic matter derived from accumulation and decay of leaves.

cambering The disturbance or arching of rock strata, usually by downslope bending or draping at the edge of its outcrop, especially on a valley side.

carr A fen or marshy area containing a variety of aquatic plants and shrubs such as alder and willow.

Cenomanian Basal stage of the Upper Cretaceous (type locality Le Mons, Cenomanum); in Britain largely represented by the Chalk Marl, Lower Chalk and Plenus Marls.

cirque A steep-walled, bowl-shaped rock-basin of glacial origin found in many mountainous areas; it may contain a small lake. Also known as a corrie, coire, combre, cwm and kar.

clasts Individual fragments of rock of variable shape, often forming part of a sedimentary rock formation.

cleavage The tendency for a rock to split or break along closely spaced parallel planes which do not correspond to the bedding plane.

col The lowest point, depression, gap or saddle in a hill or mountain range between two peaks.

colluvium A heterogeneous deposit of rock fragments, sand, silt and clay which has moved downslope under gravity.

cone-sheet A funnel-shaped arcuate zone of igneous dykes or fissures surrounding a circular or dome-shaped igneous intrusion.

conglomerate A rock composed of rounded, waterworn pebbles, cemented together in a matrix of calcium carbonate, silica or iron oxide.

crag-and-tail A resistant mass of rock ('the crag') which has withstood severe erosion by the passage of an ice-sheet while protecting softer rocks in its lee, thus creating a gently sloping 'tail'.

Cromerian An interglacial stage of the Middle Pleistocene (about 750,000 BP).

cryoplanation The process of denudation and reduction of the land surface by intensive frost action.

cuesta A ridge with a steep scarp-face (escarpment) to one side and a gentler back-slope (often a dip-slope) on the other resulting from the differential erosion of gently inclined strata.

Cuisian A stratigraphic term applied to the European Lower Eocene rocks (Tertiary); for example, the Bagshot Beds.

Danian Stage name applied to the earliest part of Palaeogene System (Palaeocene Series = Tertiary).

debris chute A gully high on a steep mountain side through which frost weathered debris falls, slides, or moves rapidly as debris flows.

Devensian The final glacial stage of the Pleistocene in Britain, succeeding the Ipswichian interglacial and lasting from about 100,000(+) to 10,000 years BP (type site, Four Ashes, near Wolverhampton, Staffordshire).

dilatation jointing A form of rock jointing in which the bedrock separates into sheets or layers roughly parallel to the rock surface; the layers increase in thickness with depth. Sheet jointing is well displayed in some granite tors (qv).

denudation chronology The study of the effects of geomorphological processes through time leading to the formation of the present landscape.

dip the inclination of a rock layer or rock surface, measured in degrees from the horizontal.

diorite A coarse-grained intrusive igneous rock, intermediate in composition between basalt and granite

dip-slope A slope the surface of which is inclined in the same direction as the dip of the underlying strata; there is rarely an exact parallelism (see also *cuesta*).

dolerite A medium grained basic igneous rock.

drift In the widest sense the term refers to all superficial deposits resting on solid bedrock; also used to describe all deposits derived from glacial erosion and deposited by ice or by meltwater (e.g. till, glacifluvial deposits); sometimes used to refer to head deposits.

drumlin A smooth hummock of glacial drift (usually till but sands and gravels may also be included), ranging in shape from a round hill to an elongated form with a pronounced long axis.

dyke(s) A wall-like body of intrusive igneous rock which cuts across bedding-planes of the surrounding (country) rock in a transgressive manner.

embayment 1. An open rounded bay forming an indentation in a coastline. 2. A large area of sedimentary rocks projecting into an exposure of crystalline (igneous) rocks.

endogenic/endogenetic factors Terms which refer to the processes that originate from within the Earth (e.g. igneous intrusions and different types of uplift and depression).

erosivity Measure of the erosive power of rainfall or runoff.

erratic A piece of rock (size varies from small fragments to blocks of great size) carried by a glacier or ice-sheet and deposited in an area of dissimilar rock types.

esker A narrow sinuous ridge of usually stratified coarse sand, gravel and boulders, usually regarded as the result of deposition from subglacial, englacial or supra-glacial melt-water streams.

eustasy World-wide change of sea-level which occurs without any local vertical movement of the land.

evapotranspiration The loss of moisture from the earth's surface by direct evaporation from vegetation, soil and water plus transpiration from vegetation.

exogenic/exogenetic factors Processes that originate outside the Earth, induced by climatic factors as well as weathering, mass movement, fluvial, glacial and marine processes.

fault A surface of fracture in bedrock representing a plane of differential movement.

feldspar Minerals consisting mainly of silicates of aluminium with those of potassium, sodium and calcium.

fetch The extent of open water over which winds generate sea waves.

fines The smaller particles, or fine fraction, identified by mechanical analysis of soils and sediments.

fissure valley A long straight valley following some line of weakness in the bedrock (such as a fault) that has been excavated by erosion.

fjärd A term applied to inlets along rocky coasts of glaciated lowland areas, which represent shallow valleys sculptured first by rivers and then by ice, and later submerged by the sea.

fjord A steep-sided glacial trough which has been partially submerged by the sea.

flags A sandy limestone or micaceous sandstone having the property of splitting easily along closely spaced parallel planes.

Flandrian An alternative name for Holocene, the post-glacial period, the last 10,000 years of the geological record; refers also to the post-glacial marine transgression.

Flandrian transgression The global (eustatic) rise of sea-level in the post-glacial (Flandrian, beginning about 10,000 years BP) period which resulted in the final severance of Britain from the continent by the creation of the Straits of Dover and the present North Sea.

fluvioglacial Associated with meltwater from ice masses; fluvioglacial deposits are those laid down by meltwater streams.

fold-axes These are the lines drawn parallel with the hinges or centre lines of folds in rock strata, from which the beds dip away in different directions to form the limbs of the folds.

foliation A layered arrangement of minerals, sometimes roughly planar, undulating or lenticular.

Foxmould Sands The lower part of the Upper Greensand.

friable A 'friable soil' is one that crumbles easily, when either wet or dry.

frost-creep When water freezes in cracks and fissures in soil and rock the resultant volume expansion brings about shattering of rocks (q.v. freeze-thaw). Alternating cycles of freezing and thawing may result in movement or creep of particles on slopes.

frost-heave The vertical lifting of the soil surface as ice-lenses expand as freezing of groundwater takes place under periglacial conditions.

Frost-sorting Repeated freezing and thawing of soils tends to sort coarser material from fines, producing polygonal networks or isolated circles of coarse material, surrounding centres of fines (*sorted polygons*). On slopes patterns become elongated by solifluction, often forming stripes of alternating coarse and fine-grained material aligned along the line of greatest slope.

gabbro A coarse-grained basic igneous rock.

Gault Clay Typically, blue silty clays of Middle and Upper Albian age with the type locality at Folkestone, Kent; name derives from the fact that it was a clay used for 'gaulting' i.e. spreading on chalk soils to improve water retention.

gelifluction Seasonal freezing and thawing of the soil and regolith which can lead to movement downslope.

geosyncline A major structural downfold in the Earth's crust, usually linear in shape.

Gipfelflur Gmn. 'summit surface': an imaginary surface that would just touch the summits of the higher peaks in an area.

glacial diffluent trough Or diffluence: a former col between adjacent valleys which has been eroded by ice-flow to such an extent that the preglacial watershed may be lowered so as to permit a post-glacial diversion of drainage by way of the new, lowered trough floor.

glacial transfluence The creation of a major new low-level gap across an original watershed as a result of erosion by ice spilling across the dividing upland.

glacial trough A valley modified by glacial erosion, which often imparts a distinctive U-shaped cross-profile and a long profile that exhibits a succession of rock steps and basins.

glacifluvial See fluvioglacial.

gneiss A coarse-grained rock with banded appearance, produced by metamorphism.

granite A coarse-grained acid igneous rock.

gravity anomaly In geophysics, the difference between a computed and an observed terrestrial gravity value.

Greensand Upper Greensand consists predominantly of green glauconitic sands of Middle to Upper Albian age, largely laterally equivalent of the Gault Clay (qv). Lower Greensand consists predominantly of glauconitic sands and clays of Aptian age.

greywacke A term originally applied to certain dark-coloured, strongly cemented coarse-grained sandstones or gritstones with large angular particles.

grid references See National Grid.

hanging valley A tributary valley to a major glacial trough where the floor of the trough lies far below the altitude of the tributary valley floor; the point of junction of the tributary is thus often marked by a waterfall.

head A poorly sorted Pleistocene periglacial deposit of rock rubble, sand and clay.

Hercynian The name given to the late Palaeozoic (Upper Carboniferous-Permian) orogeny (see table on p. 10).

Holocene The most recent of the geological periods and including all of the time since the Pleistocene, i.e. since the termination of the last glaciation (Devensian in Britain); post-glacial period beginning about 10,300 years BP.

horn A pyramidal-shaped mountain peak rising from the intersection of three or more arêtes (qv) and standing therefore between the same number of cirques.

Hoxnian The name given originally to the penultimate interglacial period in Britain and dated about 350,000 BP: the exact stratigraphic position and age of its representative sediments (at Hoxne, Suffolk) and their significance for the Pleistocene record is now in doubt.

ice-wedge A narrow crack or fissure in the ground infilled with ice.

ice-wedge polygon A type of non-sorted periglacial patterned ground restricted to the permafrost zone. Polygons are formed by intersecting surface troughs, the troughs marking the location of ice wedges which may extend downwards through the permafrost for several metres.

igneous rock Rock which has been derived from an original molten magma.

inlier An exposed mass of rocks wholly surrounded by younger ones.

interfluve An area of upland separating two river valleys; a drainage divide or watershed.

interglacial The time period (several thousand years) between two major glacial stages during the Quaternary, when the climate was milder and the ice-sheets shrank: evidence provided by plant remains indicates that some interglacials were warmer than at present.

interstadial A brief pause (may last several hundred years) between two glacial stages or two phases of glacial retreat.

involution Disturbance in the upper layers of the regolith as a result of freeze-thaw processes.

Ipswichian The last interglacial stage of the Pleistocene in Britain, characterised by a temperate climate slightly warmer than today.

isostasy A term signifying a state of equilibrium or balance in the surface crust of the Earth, the result of the necessity for equal mass to underlie equal surface area.

joint A surface of fracture in a rock mass along which there has been no differential displacement.

kame A steep-sided ridge, mound or conical hill of bedded sands and gravels, resulting from deposition of fluvioglacial material as crevasse-fillings, as ice-front deposits and especially as ice wasting takes place.

kettle-hole A circular or elongated depression in glacial drift, often with a small lake formed when a body of ice is detached, buried in sediments and subsequently melts to produce collapse of the overlying material.

Keuper Marls Part of a series of rocks of Upper Triassic age and consisting mainly of red clays and marls, with bands of more compact sandstone.

knick point A sudden break of slope or steepening of gradient along the longitudinal profile of a stream.

laccolith An intrusive mass of magma which has forced up or domed the overlying strata into a low dome or shield with a generally flat base.

Lias Term for the clays, sands and some limestones characteristic of the Lower Jurassic period in north-west Europe.

lignite Usually describes a type of low-grade coal which in structure is midway between peat and bituminous coal.

lithology The term refers to rock characteristics, especially their physical, chemical and textural composition.

'Little Ice Age' A term applied to recent advances of glaciers during the period 1550–1850 in the Alps, Norway and Iceland; also deterioriation of climate which promoted the existence of extensive snow patches and increased frost activity in upland areas in Britain during the period AD 1550–1850.

Loch Lomond Stadial A brief cold period lasting from about 11,000 to 10,000 years ago which gave rise to ice caps, valley glaciers and corrie glaciers in the uplands of Scotland, valley and corrie glaciers in N.W. England, and corrie glaciers in the Welsh mountains.

loess A fine-grained, coherent, friable and porous yellowish dust deposit associated with dust clouds carried away from newly deposited glacial and glacifluvial deposits by wind action.

London Clay Formation of (Tertiary) Eocene age found in the London and Hampshire Basins; consisting predominantly of marine clays.

Maastrichtian The latest stage of the European Cretaceous (type locality Maastricht, Holland).

machair A fine white coloured highly calcareous shell-sand, found in wide undulating tracts behind the foreshore along the coasts of Western Scotland.

Melbourn Rock The basal bed of the Middle Chalk (Cretaceous).

metamorphism A term covering all the processes by which rocks are altered in their mineralogy, texture and internal structure by heat, pressure and the introduction of new chemical substances.

mica A group of silicate minerals having the property of perfect cleavage, and splitting into thin tough lustrous plates.

Millstone Grit A hard, coarse-grained sandstone which may form massive beds, and occurs at the base of the Upper Carboniferous.

moraine An accumulation of material (it may include blocks of rock, boulders, sand and clay etc.) which has been transported and deposited by a glacier or ice-sheet.

mudstone An argillaceous (clay-rich) sedimentary rock, sometimes used to describe fine-grained shales but showing little or no lamination.

National Grid The National Grid is used on current UK Ordnance Survey maps: the axes of the grid are 2°W and 49°N, intersecting at the True Origin, but transferred to a False Origin 400 km W and 100 km N of the True Origin. The grid consists of 500 km squares, designated by a letter and 100 km squares designated by a second letter (A–Z, except I). Within these, 1 km squares and 10 km squares are accentuated. Thus, a reference can be made using two letters and six figures, on 1/50 000 maps, for example.

OD Ordnance Datum: the heights on all British maps published by the Ordnance Survey are derived from the calculations of mean sea-level at Newlyn, Cornwall (msl is calculated from hourly tidal observations).

obsequent 1. A stream flowing in a direction opposite to that of the dip of the rock strata and to the original consequent drainage, to join a subsequent (qv) stream; also, a valley carved by an obsequent stream. 2. A fault-line scarp facing in a direction opposite to that caused by the initial earth-movement as a result of differential denudation.

oolite A sedimentary rock made up essentially spherical rock particles or ooliths, formed by gradual accretion of material around an inorganic (e.g. sand grain) or organic (e.g. piece of shell) nucleus.

ord A natural shoreline feature where a combination of the reduction in height of the upper beach and the presence of migrating water channels (runnels) across the beach can lead to deeper water and hence larger waves reaching the base of the cliffs; erosion of the cliffs results from such phases of wave attack.

orogeny A major period of fold mountain formation; orogenesis is the process of fold mountain building and may include folding, faulting and thrusting (see the table on p. 10).

orthogneiss A coarse-grained, banded, crystalline rock produced by the alteration of igneous rocks by regional metamorphism.

outlier An outcrop of newer rocks surrounded by older.

overland flow Water which runs over the ground surface as sheet flow or unconcentrated wash.

ox-bow lake The surviving portion of a former river meander which has become isolated as a result of channel shifting on the flood-plain.

Palaeocene The earliest part of the Eocene, lasting from about 70 to 60 million years ago; also used for strata intermediate in age between Cretaceous and Tertiary systems and by international usage, the first of the epochs of the Tertiary period.

palsa A peaty permafrost mound typically between 1 and 7 m in height and less than 100 m in diameter, which develops in bogs and wetlands as a result of localised ice segregation due to variations in the thermal properties of the peat.

paragneiss A coarse-grained crystalline rock with a foliated texture and a banded appearance.

pegmatite A very coarse-grained igneous rock often associated with granite batholiths.

peneplain An almost level plain or undulating surface of low relief, interspersed with occasional residual hill masses (monadnocks).

pericline A crustal fold structure (anticline) in the form of an elongated dome in which the rock strata dip around a central point.

periglacial Applies to an area on the margins of an ice-sheet, and to the climate and physical processes within that area where freeze-thaw activity is paramount; also applies to the results of these processes.

permafrost Permanent freezing of the regolith and bedrock, sometimes to great depths (e.g. 600 m in Siberia); thawing may occur at and near the surface during summer, while frozen ground below still forms an impermeable layer (see periglacial).

pingo A dome-shaped or conical mound consisting of a core of massive ice covered with soil and vegetation.

pipe (piping) A sub-surface natural tunnel in soil or regolith through which water flows downslope towards stream channels, particularly well-developed in peatlands of upland Britain.

planation The term applied to the long continued denudation of rocks which has resulted in a fundamentally flat or gently sloping surface.

planation surface The result of long-continued denudation by various weathering and erosional processes leading to fundamentally flat surfaces.

Pleistocene The first epoch of the Quaternary, preceded by the Pliocene (Tertiary) and succeeded by the Holocene: used to cover different time periods by various authorities, usually about 2.4 million years.

plucking The pulling away or quarrying of masses of rock by glacier ice by ice freezing on to projecting parts of the rock surface and then detaching them as the glacier moves.

plug A more or less cylindrical mass of solidified lava, occupying the vent of a dormant or extinct volcano, now exposed by denudation (e.g. Castle Rock, Edinburgh).

plutonic rock An intrusive igneous mass which has cooled slowly at great depth in the Earth's crust, resulting in coarse-grained texture (e.g. diorite, gabbro and granite).

pneumatolysis Chemical changes produced in rocks by heated gases and vapours other than water during igneous activity.

podsol A soil formed under cool, moist climatic conditions with a vegetation of coniferous forest or heath, especially from sand parent materials, resulting in intense leaching of base-salts and iron compounds.

point bar deposits Alluvium, sand and gravel deposited on the shelving slip-off slope of the inside bends of meandering river channels.

pollen zones Zones which display different fossil pollen assemblages and thus different plant assemblages; from these data past climatic changes can be inferred.

pool and riffle A pattern of alternating deeper pools and shallow gravel bars (riffles), which develops in many streams.

pore water pressure The pressure exerted on the regolith and rock by water contained within the interstices, resulting in landslips and mud slides.

Pre-Boreal A climatic period marking the beginning of the Holocene and dating from about 10,300 to 9,500 years BP; characterised by dry, cold conditions, with a birch–pine flora: Pollen Zone IV in Britain.

Primary Era An old term for Pre-Cambrian time which eventually became used to mean the whole of the Palaeozoic era (see the table on p. 10).

proglacial lake A term used for an area of water dammed between the edge of an ice-sheet (or glacier) and a topographical divide.

protalus rampart A ridge consisting of an accumulation of coarse angular debris, resulting from the sliding of frost-riven debris across perennial snow-banks.

Purbeck Limestone A general name to include the 'Purbeck Beds' of Jurassic and Cretaceous age.

Pyramidal peak See horn.

quartzite A hard resistant rock consisting largely of cemented quartz grains.

radiocarbon dating Carbon 14 is a radioactive isotope of carbon with a half-life of 5,570 years. Created in the upper atmosphere it is absorbed by living matter but after death of an organism, or burial in sediments, 14C ceases to be assimilated and the content decays at a known rate. By measuring the proportion of 14C to the total carbon content of, for example, a buried piece of wood, charcoal, bone or peat the age of the fossil item can be calculated. It can be used to measure ages up to 50,000, sometimes 70,000 years BP.

recessional moraine A terminal moraine which is the result of a brief pause in the retreat, or even a slight readvance, of an ice sheet or valley glacier.

reef knoll A dome-like mass of limestone that has grown upwards from a reef in order to keep pace with the deposition of surrounding sediments.

regolith Mantle or covering of more or less disintegrated, loose unsorted rock waste overlying the bedrock, including the soil and any superficial deposits of alluvium, glacial drift, loess and peat, etc.

relict landform A landform resulting from the action of processes which are no longer operative upon it.

Rhaetic The old name for the Penarth Group, a series of mainly shallow marine sediments of latest Triassic age, transitional to the overlying Jurassic rocks: usually consists of a series of shales, marls and limestones (see Westbury Shales).

rhyolite A fine-grained acid extrusive volcanic rock having more quartz and fewer ferro-magnesian minerals than granite.

ria A submerged coastal valley or estuary.

riegel See rock bar

roche mountonnée A glacially-moulded mass of rock having a smooth, gently sloping rounded upstream side (abraded and bearing frequent striae) and a steep, roughened and often broken downstream side (the result of plucking): common in glaciated areas of Britain and varying in size from a few metres in dimensions to rocky hills tens of metres high and hundreds of metres in length.

rock-bar Bedrock threshold marking the exit of a cirque or separating one rock basin from another along the course of a glacial trough.

Rogen moraine A glacial moraine formed in subglacial positions transverse to the predominant ice flow direction.

rotational slip A downslope movement *en masse* of solid material such as a rock or ice-mass on a well-defined concave upwards slip-plane, leaving the slipped mass with a pronounced back-slope facing upslope (e.g. the landslips of The Warren near Folkestone).

runoff The fraction of precipitation that leaves drainage basins as river discharge. Runoff amounts vary with annual rainfall, soil and rock permeability, topography, evapotranspiration rates and vegetation cover.

saltation The process by which sand, gravel and boulders are moved along the bed of a river in hops.

sandur An outwash plain composed of sand, silt and gravel, representing glacifluvial sediments carried away from the front of an ice-sheet by meltwater streams.

sarsen Blocks of siliceous sandstone found mainly on the Chalk Downs and representing remnants of a now vanished cover of early Tertiary cover rocks: often concentrated on valleysides and in valley bottoms having been moved downslope by periglacial solifluction (e.g. in Wiltshire).

schist A medium to coarse-grained rock, usually foliated as a result of metamorphism.

Senonian Represented by the Upper Chalk (Cretaceous) in Britain.

shale An argillaceous sedimentary rock formed from deposits of fine mud, compacted and compressed to such an extent as to develop thin laminations (qv mudstone).

shatter-belt A narrow tract of broken crustal rocks where earth movements (along fault zones) have created a distinct line of weakness which can be exploited by weathering and erosion.

sheet jointing See dilatation jointing.

sill A near-horizontal or tabular sheet of intrusive igneous rock intruded between the bedding planes of the surrounding (country) rock (e.g. Great Whin Sill).

sink hole A funnel-shaped depression (or sink) in limestone country; formed by surface solution or by subterranean collapse of a cave system: surface streams often disappear underground via sink holes.

sinuosity ratio A term which refers to the degree of meandering exhibited by a stream channel: expressed either as the ratio between channel length and valley length, or as the ratio between channel length and the length of the axis of the meander belt.

slate A fine-grained metamorphic rock formed from deposits of mud or clay which have been subjected to great pressure as a result of earth movements; splits easily into thin layers along the cleavage: also formed from fine-grained volcanic ash.

slumping Mass movement involving the shearing of rocks and the tearing away of masses of material as blocks, with a distinct rotational movement on a curved, concave-upwards plane (e.g. along the S. coast of the Isle of Wight).

solifluction Slow downslope flow-like movement of the soil due to spring thawing of soil ice, which releases excess water and causes considerable reduction in soil strength. Frost creep may also contribute to the total movement of the soil. Over several thousands or tens of thousands of years, large volumes of near-surface sediments may be moved downslope by this process.

sorted polygons A type of periglacial patterned ground consisting of a polygonal network of coarse clasts which surrounds finer-grained polygon centres.

Sparnacian Uppermost stage of the European Palaeocene Series above Thanetian (see Tertiary).

spilite A basic igneous rock akin to basalt and commonly found interbedded with marine sediments.

stadial A short pause or minor readvance, sometimes lasting up to a few thousand years, during the decline of an ice-sheet or glacier, often marked by a *stadial moraine*. (e.g. Loch Lomond Readvance or Stadial).

storm surges A rapid rise in the level of the sea above a predicted tidal height.

striae Scratches and grooves found on ice-scoured rock surfaces, produced by the dragging of rock fragments at the base of moving glaciers or ice sheets.

strike In the geological sense, the direction of a horizontal line drawn through any point on a dipping rock surface or bedding plane.

syncline A downfold in the rocks of the Earth's crust, with strata dipping inward towards a central axis, in contrast to an anticline.

synclinorium A complex syncline of considerable lateral extent where many minor downfolds and upfolds occur.

sub-aerial A term used to refer to processes operating and features forming on the surface of the Earth, as distinct from submarine, sub-aqueous or subterranean phenomena.

Sub-Atlantic The climatic period since 3,200 BP, marked by a deterioration of summer temperatures and a return to milder and moister conditions: an alder, oak, elm, birch and beech flora gradually spread with beech (and hornbeam in S.) dominant: regeneration of peat bog growth. Pollen Zone VIII in Britain.

Sub-Boreal A climatic period dating from about 5,200 to 3,200 BP, with cooler but rather drier conditions than in the previous Atlantic: ash, birch and pine forests slowly assumed an ascendency over the oak forests: Pollen Zone VIIb in Britain.

subglacial stream A stream of melt water flowing beneath a glacier or ice sheet in a tunnel.

submerged forest A layer of peat or peaty-mud with embedded tree-stumps, some in the position of growth, occurring between tide-levels or below present low tide: indicative of the drowning of a former land surface by a rise of sea-level.

subsequent stream A stream which has developed by headward erosion along lines of geological weakness, such as outcrops of clays, shales and weak sandstones, fault-lines, major joints and anticlinal axes.

talus Slope deposits of mainly angular debris lying at the foot of a steep bedrock face from which the fragments have fallen, following the action of various weathering processes. *Scree* is an alternative name.

tectonic activity A term applied to all the internal forces which deform the Earth's crust and affect landforms related to folding, faulting and warping, while influencing patterns of sedimentation.

terrace A relatively flat shelf or bench along the side of a valley (e.g. *river terrace*,) where it represents part of the former floodplain, now dissected as a result of downcutting by the river.

Tertiary Time period comprising the Palaeogene and Neogene systems of the Cainozoic Era. Generally taken to exclude the Pleistocene and Holocene periods (Quaternary).

thalweg A term used to denote the longitudinal profile of a river or river valley. Also used to describe the course of maximum water velocity in a river channel: this tends to meander even in straight river channels.

tholeiite A type of basalt.

thrust A compressional force exerted on rock strata in a low-angle or near-horizontal plane.

till Usually an unsorted, heterogeneous assortment of rocks, clay, silt and sand, carried within, upon or near the base of a glacier or ice-sheet.

tor A prominent often castellated rock mass standing tens of metres above the general surface of the surrounding area (e.g. granite tors of Dartmoor).

trachyte A fine-grained extrusive intermediate igneous rock.

trough-end The abrupt upper end of a glacial trough, above which cirques, formerly feeding ice into the trough, hang above the floor.

tuff A rock of volcanic origin, composed of mainly medium- to fine-grained (particles < 4 mm in diameter) particles formed from a compacted or cemented mass of volcanic ash.

unconformity A surface separating one set of rocks from another, the overlying set of rocks possessing a different structural arrangement.

undercliff A term describing the masses of material which can accumulate at the base of a cliff (marine) as a result of falls due to rock weathering.

valley bulging A term given to the up-arching of an incompetent rock bed, which is thrust upwards into the floor of a valley by the weight of the overlying rocks forming the flanking valley sides and adjacent uplands.

valley step A sudden steepening in the long profile of a glacial trough (see rock bar) between successive rock basins.

warm-based glacier A glacier at the base of which the temperature of the ice is at pressure melting point. The basal ice is thus not frozen to its bed and is frequently separated from the bed by a thin film of meltwater and the ice may therefore slide over its bed.

watershed breaching by ice See glacial transfluence.

wave climate The spectrum of sea wave conditions, that is, wave height, length, strength and orientation that characteristically arrive at a particular coastal sector, unit or landform.

Weald Clay The Weald Clay Formation is the upper part of the Wealden Group. In the type area of the Weald it attains a maximum thickness of c.450 m. It is of Lower Cretaceous age and represents the Hauterivian and Barrenian stages of the Cretaceous.

weathering The disintegration and decay of rock *in situ* producing a mantle (regolith) of waste material.

Westbury shales Now known as the Westbury Formation, a rock formation in the lower part of the (Jurassic) Penarth Group (see Rhaetic): consists normally of black shales with one or more bone beds.

Wolstonian A glacial period of the Middle Pleistocene (type site for sediments at Wolston, Warwickshire): exact stratigraphic position and significance of these sediments under review: previously dated approximately from 350,000 to 130,000 BP and believed terminated by the Ipswichian interglacial.

wrench fault A nearby vertical strike-slip fault with a large horizontal displacement.

Ypresian The lowermost stage of the European (Tertiary) Eocene Series (above Sparnacian qv)., for example, the London Clay in Britain.

The glossary has been compiled using a variety of sources, including especially J. B. Whittow, *The Penguin Dictionary of Physical Geography*, Harmondsworth, 1984; D. G. A. Whitten and J. R. V. Brooks, *The Penguin Dictionary of Geology*, Harmondsworth, 1977; and F. J. Monkhouse and John Small, *A Dictionary of the Natural Environment*, London, 1978. Grateful thanks is accorded to Dr J. C. W. Cope for help with some geological definitions and to Mrs D. J. Nuttall for preparing the manuscript.

Index

Photographs are indicated by page numbers in italics.